LEARNING DIGITAL ELECTRONICS THROUGH EXPERIMENTS

EDWARD J. PASAHOW

D0222106

Gregg Division
McGraw-Hill Book Company

New York
Atlanta
Dallas
St. Louis
San Francisco
Auckland
Bogotá
Guatemala

Hamburg
Johannesburg
Lisbon
London
Madrid
Mexico
Montreal
New Delhi

Panama
Paris
San Juan
São Paulo
Singapore
Sydney
Tokyo
Toronto

For Geneva and Stan

Library of Congress Cataloging in Publication Data

Pasahow, Edward (date)
    Learning Digital Electronics Through Experiments.

    Includes index.
    1.   Digital electronics—Experiments.   I.   Title.
TK7868.D5P37   621.381   81-2688
ISBN 0-07-048722-7   AACR2

2 3 4 5 6 7 8 9 0   DODO 8 9 8 7 6 5 4 3 2

Sponsoring Editors: Mark Haas and Gordon Rockmaker
Editing Supervisor: Evelyn Belov
Design Supervisor: Nancy Axelrod
Production Supervisor: Priscilla Taguer

Cover Photographer: Ampersand Studio

ISBN 0-07-048722-7

Contents

Preface

Learning Digital Electronics Through Experiments is precisely that—a book designed to develop your ability to use integrated circuits (ICs) in practical projects. Previous experience with electronics is not necessary in order to understand the examples. However, if you are familiar with the basic concepts, you will recognize applications throughout the book of what you already know. The entire discussion is based on the popular 7400 transistor-transistor logic (TTL)—including the low-power Schottky series—and the 4000 complementary metal-oxide semiconductor (CMOS) series integrated circuits. The actual ICs illustrate the principles, so a great many useful circuits are provided and the "reality" of the hardware in each chapter is apparent.

Treating the ICs as modular building blocks demonstrates how easily they can be used. Hands-on experiments emphasize the low cost, simplicity, and effectiveness of digital devices.

Each chapter presents a set of ICs which perform specific function. Because each chapter can stand alone as a complete discussion of its subject, you need not follow the chapters in consecutive order. Instead, simply turn to the subjects that interest you. Each experiment you perform will give you that much more experience for the next one—and the next.

The illustrations supplement the text by summarizing key points. In fact, the illustrations form a kind of visual narrative, easy to comprehend even without reading the text. Viewed separately, the figures provide a capsule review of the principles discussed.

With this book and a minimum of equipment, you—the technician, engineer, or hobbyist—will be able to teach yourself IC technology. After you complete each topic, you will be able to construct, understand, and troubleshoot the digital circuits involved.

Edward J. Pasahow

1
Introduction to Integrated Circuits

Because you have picked up this book, you are obviously interested in digital circuitry. With this book and some low-cost components, you will learn how the circuits work, what they can do, and how they can be combined to do an entire job.

More and more electronics are becoming digital. Clocks and watches were entirely analog a decade ago, yet today new timepieces that are not at least partially digital are a rarity. The gauges and the meters on the dashboards of cars now measure mileage, speed, gasoline supply, and temperatures with numerical readouts. Even the door locks on cars are being replaced with digital combination locks. Almost every home has a pocket calculator, and many also have microwave ovens, television sets, and washing machines that are operated through a digital control panel or a keyboard. Many other examples such as telephones, personal computers, video disks, games, and even the automatic tellers at the bank reinforce the conclusion that digital systems will be used in an ever-increasing segment of the machines around us.

The purpose of this book is to introduce you to the functional elements that make up these electronic marvels. Once you grasp the underlying principles, you will come to realize that digital systems are quite simple in concept. Their complexity comes about because they are capable of rapidly repeating the same action, which, in human terms, is amazing.

SUGGESTIONS FOR USING THIS BOOK

By this time you are probably wondering, "How much do I have to know before I can understand digital electronics?" Surprisingly, very little previous knowledge of electronics is required to learn about digital integrated circuits. Naturally, if you already have considerable electronics background, you will grasp the significance of some ideas more quickly and will see some of the more subtle concepts more easily. However, a beginner who knows Ohm's law and a little about solid-state diodes will find that proceeding from that understanding through this book is quite practical.

Another question that comes to mind is, "How many experiments do I need to do?" It goes without saying that the more you are able to complete, the better you will know integrated circuit operation. To discover the answer to this question, you should read through all the experiments in a chapter one time before starting any of them and then select the ones that seem to be unclear from the first reading as the most important for you to work through. The later experiments build on experience gained in earlier ones, so you may find that after the experiments in the first half of the chapter have been completed, the later ones need not be worked to understand the material. Another reason for choosing not to work an experiment may be that you lack the equipment. For example, in two or three experiments, an oscilloscope is needed. If you do not have access to that test equipment, it will not be possible to run the experiment.

Every effort has been made to avoid the requirement for expensive test equipment in this book. In fact, the experiments in this chapter detail how to build your own.

THE SCOPE OF DIGITAL INTEGRATED CIRCUITS

To put this subject in perspective, let us consider what types of digital integrated circuits (ICs) there are and how they relate to one another. Figure 1-1 gives one view of this relationship. While it is difficult to picture all the possible ties between the different types of circuits, the illustration does show that logic circuits, also called *gates,* form the basis for the more advanced forms of digital components. These gates are able to combine their inputs in various ways to produce an output that depends on the present values at the input terminals.

At the next level of the pyramid, several gates are interconnected to act as flip-flops which can hold their output value constant even after the inputs are removed. Such a circuit is said to have a memory. By rearranging the logic gate in a different fashion, you can build circuits that perform addition, subtraction, multiplication, and other arithmetic operations.

Just as flip-flops were composed of logic gates, some third-level pyramid components are formed from flip-flops. Both counters and registers are built using several flip-flops in each circuit. Similarly, several arithmetic circuits can be packaged as an arithmetic logic unit (ALU).

Continuing our climb upward, memories for equipment such as computers or

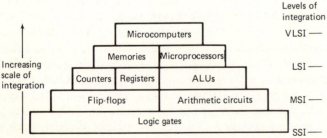

Fig. 1-1 One view of the world of digital electronics. In moving up the pyramid, the scale of integration increases.

calculators can be thought of as organized collections of registers. A memory the size of your little finger can store hundreds of thousands of pieces of information. Another very powerful IC is the microprocessor. These ICs contain the essential control, arithmetic, and decision-making components of a computer. Possibly the most sophisticated of all are the microcomputers. A single IC provides the processor, memory, and input/output functions for an entire computer.

As you move up the pyramid, you will see that individual components are involved at each layer. The number of components in a single circuit is a measure of the level of integration. At the lowest levels, logic gates are examples of small-scale integration (SSI). When there are more than 10 gates per package (reached at the second level), medium-scale integration (MSI) is involved. By the fourth tier the circuits require large-scale integration (LSI), more than 100 gates in one package. At the highest levels, thousands of gates are needed for very-large-scale integration (VLSI).

CONVENTIONS

Digital electronics are characterized by the use of two-state signals exclusively. Throughout this book reference will be made to two voltage levels: high and low. A high voltage will be defined as one in the range of 2.4 to 5 volts (V). A low voltage is between 0 and 0.8 V (although in some cases a slightly negative value will be allowed for a low-level voltage, more positive than -1.5 V, for example). All other voltages are not allowed as inputs to these circuits.

Because only two ranges of voltages are permitted, it becomes convenient to associate a symbol with each level. Often the capital letters H and L will represent the high and low levels, respectively. Sometimes the level is of no consequence. If it does not matter whether the voltage is high or low, the value is a *don't care* condition. The letter X is frequently used to indicate a don't care level for a signal. Another frequently used technique is to let the digits 0 and 1 stand for the two levels.

These two digits form a special number system called *binary*. This number system will be discussed in further detail in Chap. 6. For now, it is sufficient to observe that the digit zero can mean either a high or a low level; the choice is arbitrary. If zero means a low level, a *positive logic* convention is in use. In that case, a one stands for a high level. The opposite situation, one meaning low levels and zero meaning high, is an application of *negative logic*. Both of these techniques are in use. This subject is discussed in further detail in the experiments which follow.

Another procedure followed in this book is the exclusive use of *conventional* current. The effect of this choice is that current will be represented as flowing from the more positive to the more negative potential. Figure 1-2 provides several examples. The current flows from the positive terminal of the battery in Fig. 1-2*a* through the resistor to ground. With a diode in the circuit, the diode symbol points in the direction of forward bias current flow. If two batteries are connected so as to oppose each other (Fig. 1-2*c*), current still flows from the more positive (5-V) terminal to the less positive (3-V). You may be more familiar with *electron* current, which flows in the direction that electrons move. The electron current direction is simply the opposite of conventional current flow.

In the experiments, you will find that high-level inputs cause current to flow *into*

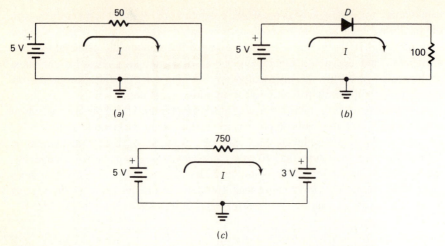

Fig. 1-2 **With conventional current, the flow is from the more positive to the more negative potential.** *(a)* **Resistor circuit;** *(b)* **diode;** *(c)* **two opposing batteries.**

the IC, thus requiring a current source on that terminal. A low-level input causes current to flow *out*, meaning that the terminal with the low level must be attached to a current *sink*. For convenience, a system for referring to current flow involving ICs has also been defined. If the current flows *into* the device, it is a positive current. Current flow *out* is negative.

A NOTE ON EQUIPMENT

For these experiments one of the most important items will be a breadboard socket (see Fig. 1-3). These sockets come in various sizes and make connections without soldering. The largest size that your budget allows is recommended, because you will find that you want to build bigger and bigger circuits as your confidence grows. For these experiments, a socket 2 × 6 inches (in) [5.1 × 15.2 centimeters (cm)] is adequate. To remove ICs from the socket without bending the pins, pry them out with a small screwdriver in the center channel that runs under the IC.

The ICs themselves will be your most frequent purchase. A variety of them is called for. The types used in the experiments cost considerably less than a dollar each in a plastic package, and hobby suppliers often sell them at bargain prices.

Standard tools and hook-up wire will be used routinely. A suggested starter set would include diagonal pliers, needle-nose pliers, a knife, a wire stripper, a small screwdriver, a miniature soldering iron, and a voltmeter.

In working with ICs, a good low-impedance power supply is necessary. A power supply suitable for all experiments in this book, as well as many transistor-transistor logic (TTL) applications, requires regulation to within 250 millivolts (mV) with a load-carrying capacity of at least 500 milliamperes (mA). Instructions for building such a power supply follow in this chapter.

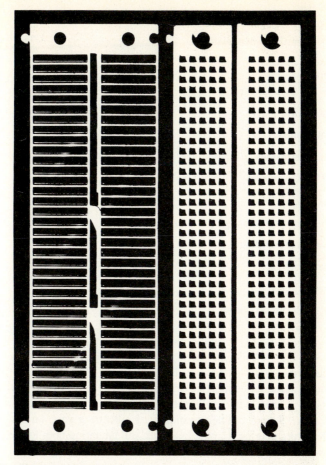

Fig. 1-3 Breadboard sockets make the assembly of experimental circuits simple and quick. No soldering is necessary.

Other test equipment needed for IC experimentation may be bought, or you can make your own much less expensively. Of course, commercial models are likely to offer more convenience features. In this chapter you will learn how to build a logic probe, a clock generator, a pulser, and a simple decade counter. Even if you already own these items, making your own will show you how they work and also give a sense of accomplishment. The circuit may also be useful as a building block in other digital equipment. Almost all of them will need power supplies, indicators, and clocks.

EXPERIMENT 1-1   BUILDING A POWER SUPPLY

Purpose

To construct a regulated, 1-ampere (A) power supply suitable for use with the digital experiments in this book.

Parts

Item	Quantity
Power transformer [120-V primary, 6.3-V, 1-A secondary]	1
Power cord	1
IN4001 diodes	4
2200 microfarads (μF), 16-V electrolytic capacitor	2
LM309K regulator	1
Heat sink for TO-3	1
Perf board for mounting (3 $\times$ 4 in)	1
4-contact terminal strip	1
Screws and nuts	5
Silicone grease	
Plastic electrician's tape	

Procedure

Step 1. The parts layout is shown in Fig. 1-4. Place the parts on the perf board. Mark the points where holes must be drilled to attach the transformer, the terminal strip, and the regulator. After drilling the holes, mount the transformer with two screws and nuts. Spread a liberal amount of silicone grease on the bottom of the LM309K before placing it in the heat sink and mounting both to the board with the two screws. Attach the terminal strip with the remaining screw and nut. Mount the remaining parts on the board by slightly bending the leads to hold them in place.

Step 2. Following the schematic of Fig. 1-5, wire the circuit as described in the remaining steps.

Step 3. Solder the power cord lead to the primary leads of the transformer. Wrap any bare wires with tape.

Step 4. Twist one lead from the transformer secondary together with the anode (end with no band) of $D1$ and cathode (banded) of $D3$. Solder.

Step 5. Twist the other lead from the transformer secondary to the cathode (banded) of $D4$ and anode (no band) of $D2$. Solder.

Step 6. Twist the cathodes (banded) of $D1$ and $D2$ together with the positive terminals of the two capacitors. Attach a 3-in [7.6-cm] wire to the same point. Solder.

Step 7. Run the wire to the input of the LM309K. Solder.

Step 8. Solder another wire to the output of the LM309K. Run it to the terminal strip. This is the 5-V pin for the power supply.

6

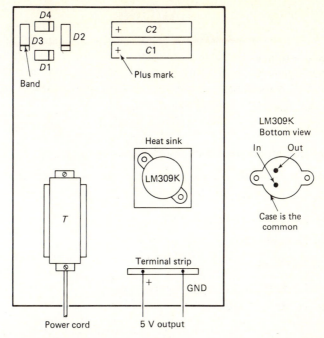

Fig. 1-4 Parts placement for the power supply. Carefully note the orientation of the diodes and the capacitors.

Step 9. Twist the anodes (unbanded) of *D*3 and *D*4 together. Attach a wire between that point and the unused leads on the capacitors. Also, from the same leads of the capacitors, attach a 3-in [7.6-cm] wire. Solder all.

Step 10. Run the other end of that wire to the case of the LM309K. Attach with another 2-in [5.1-cm] wire to the case. (Attach this wire by slightly loosening one screw, and then tighten.)

Step 11. Solder the remaining end of the 2-in wire to the terminal strip. This is the ground for the power supply.

Step 12. Check all your wiring. Clip off ends of any wires that protrude.

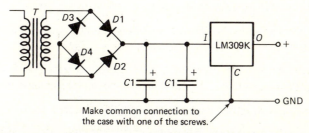

Fig. 1-5 Schematic for the power supply showing the stepdown transformer, the rectifying bridge, filtering capacitors, and the voltage regulator.

7

CAUTION

Step 13. Plug the supply into the wall socket.

Step 14. Test the output voltage with a direct-current (dc) voltmeter. You should read between 4.5 and 5.5 V.

Ins and Outs

The power supply operation is based on a series of basic electrical and electronic operations. The transformer steps the high voltage down to 6.3 V. This sinusoidal wave is rectified by the full-wave rectifier formed by $D1$ through $D4$. The two capacitors filter the voltage which is regulated by the LM309K.

Conclusions

By connecting a power transformer, a full-wave rectifier, filtering capacitors, and a voltage regulator, a 5-V dc power supply can be built.

EXPERIMENT 1-2 CONSTRUCTING A LOGIC PROBE

Purpose

To make a test probe that indicates whether a logic level is high or low.

Parts

Item	Quantity
Light-emitting diode (LED)	1
330-ohm (Ω) resistor	1
Alligator clip	1

Procedure

Step 1. Wire the circuit as shown in Fig. 1-6. You will probably want to use the probe over and over again, so soldering the components together is suggested. A convenient way to mount the probe is to place it in an old ballpoint pen barrel.

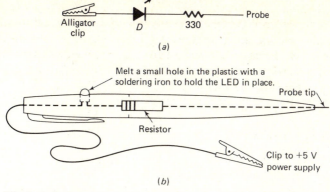

Fig. 1-6 The logic probe circuit indicates high or low levels. Mounting the probe in a ball-point pen barrel creates a handy piece of test equipment. *(a)* Schematic; *(b)* mounted in a ball-point pen barrel.

Step 2. Test the probe by connecting the alligator clip to 5 V. Touch the probe to ground. The LED should light.

Step 3. Now touch the probe tip to 5 V. The LED remains off.

Ins and Outs

The LED operates much like an ordinary diode in regard to current flow. When the anode is more positive than the cathode, current flows and the LED is illuminated. Therefore, when the probe is touched to a low (ground) signal, the LED is on. When the LED is off, the probe indicates that the input signal is high. If the input is a train of alternately high and low pulses, the LED will blink if the frequency is less than 25 hertz (Hz). At higher frequencies, the LED glows, but with less intensity than that produced by a high input.

Conclusions

A logic probe indicates whether a signal is high or low. The one you built in this experiment is suitable for testing digital circuits which have a power supply voltage of 5 V. The probe is clipped to the power supply output and then touched to one of the pins of an IC. If the pin is low, the LED glows. A high signal level on the pin causes the LED to remain off.

EXPERIMENT 1-3   PRODUCING A SQUARE-WAVE SIGNAL

Purpose

To make a square wave, or clock, generator with a variable frequency for the pulse train.

9

Parts

Item	Quantity
555 timer	1
8-pin IC socket	1
15-kilohm (kΩ) resistor	1
68-kΩ resistor	1
62-picofarad (pF) mica capacitor	1
0.001-μF disk capacitor	1
10-μF tantalum capacitor	1
Small perf board for mounting	1
Oscilloscope	1

NOTE: Working dc voltage on all capacitors is to be equal to or greater than 16 V.

Procedure

Step 1. Wire the circuit as shown in Fig. 1-7. Solder all connections. Do not put the 555 in the socket until all soldering is completed. The output frequency of the generator depends on the value of the capacitor. (This frequency is not exact because of the tolerance of the components.)

Step 2. Insert the 62-pF capacitor. Using the oscilloscope, measure the period of one cycle. [*About 10 microseconds (μs)*]

Step 3. Change the value of the capacitor to 0.001 μF. Again measure the output period. [*About 1 millisecond (ms)*]

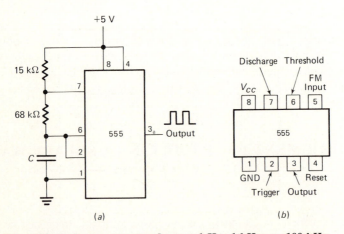

Fig. 1-7 The 555 clock generator produces a 1-Hz, 1-kHz, or 100-kHz square wave depending on the value of the capacitor chosen.

Step 4. Finally use the 10-μF value for the capacitor and measure the period. [*About 1 second (s)*]

Ins and Outs

The 555 timer can be connected in a number of ways. Here you are using it as an astable multivibrator. The frequencies of the output are listed in Table 1-1.

Table 1-1 Generator Output Frequency And Period

Value of Capacitor	Frequency	Period
62 pF	100 kHz	10 μs
0.001 μF	1 kHz	1 ms
10 μF	1 Hz	1 s

The components, as used in this experiment, are shown in Fig. 1-8. The key elements of the 555 are the two analog comparators. The comparators are able to measure two input voltages and produce an output that indicates which is higher. The flip-flop output changes when comparator 1 or 2 is active. That flip-flop output will be either high or low. The transition between levels produces a square wave. The frequency of the waveform is controlled by the size of the capacitor. A larger capacitor takes longer to charge; hence the frequency is lower.

Conclusions

The 555 timer provides the basis for a square-wave generator that uses only two resistors and a capacitor. The frequency of the output depends on the size of the capacitor used, if the resistor values do not change.

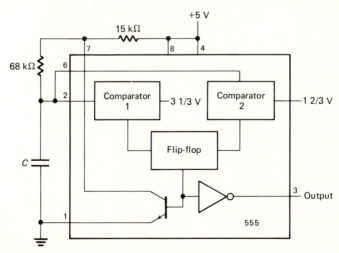

Fig. 1-8 The internal components of the 555 consist of two analog comparators, a flip-flop, an inverter, and a discharge transistor.

EXPERIMENT 1-4   CONSTRUCTING A LOGIC PULSER

Purpose

To build a momentary contact logic switch which produces a pulse signal.

Parts

Item	Quantity
7400 IC	1
Single-pole double-throw (SPDT) slide switch (spring return)	2
900-Ω resistor	2
14-pin IC socket	1
Small perf board for mounting	1

Procedure

Step 1.   Wire the circuit as shown in Fig. 1-9. Do not put the IC in the socket until you have soldered all the connections.

Step 2.   Connect the power supply and ground leads.

Step 3.   With the logic probe, test the two outputs. Which one is high? (*Pin 3*)

Step 4.   Now hold the switch in the on position. Repeat the test. (*Pin 6 is high*)

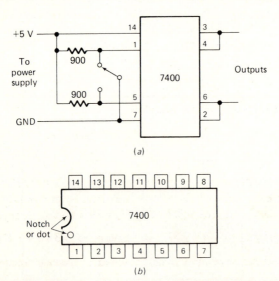

Fig. 1-9   The logic pulser produces crisp square waves. Note the pin numbering convention for the IC. (*a*) Schematic; (*b*) IC pin numbering.

Ins and Outs

The logic pulser is basically a debounced switch. The operation of both the 7400 and this circuit are explained in later chapters. Without the IC, the output of the pulser would be a very ragged waveform. Adding the IC to the switch produces crisp square waves ideally suited for testing digital circuits.

Conclusions

The pulser has two outputs that are always in opposite states regardless of switch position. The output that is best suited for testing the circuit can be chosen. In this way the test can start using either a high- or a low-level input depending on whether pin 3 or 6 is used. Throwing the switch the other way forces the signal to the opposite level.

EXPERIMENT 1-5   COUNTING PULSES

Purpose

To build a single stage decade counter.

Parts

Item	Quantity
7447 IC	1
7490 IC	1
330-Ω resistor	7
IEE 1712 seven-segment LED display	1
16-pin IC socket	1
14-pin IC socket	2
Small perf board for mounting	1

Procedure

Step 1.   Solder the circuit shown in Fig. 1-10. Do not insert the ICs or the displays until all soldering is completed. (Refer to Appendix A for pin numbering on the ICs.) As Fig. 1-10*b* shows, the IEE 1712 does not have a normal arrangement of pins.

Step 2.   Set the clock generator built in Experiment 1-3 for a 1-Hz frequency.

Step 3.   Connect the power and ground leads of the decade counter.

Step 4.   You should observe the display counting from zero to nine, and then return to zero and repeat.

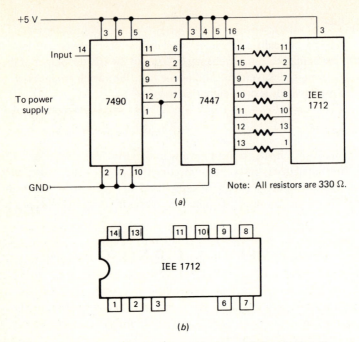

Fig. 1-10   The decade counter accumulates 10 inputs and then begins the counting sequence again. *(a)* Schematic; *(b)* IEE 1712 display pin assignments.

Ins and Outs

The decade counter is comprised of two ICs and the display. The 7490 performs the counting function. It starts with zero and adds one to the total as each input pulse is received. When the count rolls over to zero again, 10 input pulses have been received. You may wish to compare the accuracy of the 1-Hz 555 clock generator to your watch. Time 10, 20, and 30 counts. More than likely you will find that the 555 frequency is only an approximation of 1 Hz.

The 7447 converts (decodes) the count sent by the 7490 to the signals needed to light each of the decimal digits on the LED; hence the 7447 is called a *decoder/ driver*. These ICs and the display are examined in the upcoming experiments.

Conclusions

The decade counter is a convenient item for totaling the number of input pulses that have arrived. It consists of a decade counter, decoder/driver, and a seven-segment LED display.

2

Getting Started

General concepts and simple models are applied over and over again in work with ICs. In this chapter you will learn how to use the ''basic tool kit'' of IC analysis. These first investigations will allow you to actually see the operation of circuit elements that implement the inverter, AND, and OR functions that are used in all logic design. Then a discussion of how to use the IC forms of these functions follows. From the start you will soon be putting the functions together to realize the truth tables for specific expressions and investigating how to minimize the number of gates necessary. Finally, you will build circuits using combinations of simpler functions— the foundation of an understanding of how IC modules can be combined into a complete digital equipment item.

EXPERIMENT 2-1   VISUALIZING AN INVERTER

Purpose

To demonstrate the operation of a diode inverter.

Parts

Item	Quantity
100-Ω potentiometer	1
LEDs	3
130-Ω resistor	1
Voltmeter	1

Procedure

Step 1.   Wire the circuit shown in Fig. 2-1. The voltmeter is not to be connected at this time. Rotate the potentiometer shaft so that the minimum voltage is applied at point A. Measure the voltage at that point. Is diode $D1$ on or off? You should have measured about 0 V and found that the LED is on.

Step 2.   Measure the voltage at point B. Are diodes $D2$ and $D3$ on or off? *(Both diodes should be off.)*

Step 3.   Now rotate the potentiometer all the way in the opposite direction. What is the voltage point at A? What is the condition of $D1$? Do you find that $D1$ is off?

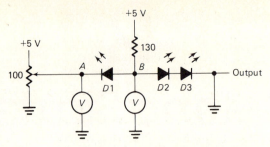

Fig. 2-1 The diode inverter causes the output level at point *B* to be opposite that at point *A* as shown by the voltmeter readings and the LEDS.

Step 4.   Check the voltage at point *B*. Are *D*2 and *D*3 on or off now? *(Both on)*

Ins and Outs

The inverter causes the output to be the *inverse*, or the *complement*, of the input. Your readings could be arranged in a list like that shown in Table 2-1.

Table 2-1   Inverter Inputs and Outputs

Input Voltage, V	*D*1	Approximate Voltage at Point *B*, V	*D*2	*D*3
0	On	1.7	Off	Off
5	Off	3.4	On	On

As you can see for an input of about 0 V, diode *D*1 is forward-biased from the 5-V power supply through the 130-Ω resistor. No current flows through the output diodes. There is no current flow through the two series LEDs because it takes 1.7 V to forward bias each diode. Diode *D*1 is forward-biased, and point *B* is at approximately 1.7 V. This voltage is insufficient to cause current to flow through the two diodes in series, which would require 3.4 V.

When the input voltage is 5 V, diode *D*1 is reverse-biased, and thus does not glow. The power supply will now cause a current to flow in the leg with *D*2 and *D*3. The diodes are lighted and the voltmeter reads 3.4 V at point *B*.

The symbol for an inverter is a triangle with a small circle on the output terminal (see Fig. 2-2). The symbol implies the voltages shown in Table 2-2. A "high" (H) voltage is any value between 3.0 and 5.0 V, and a "low" (L) voltage must be in the range of 0–1.7 V.

Table 2-2   Inverter Voltage Table

Input	Output
Low (L)	High (H)
High (H)	Low (L)

A high input produces a low output.

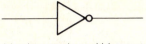

A low input produces a high output.

Fig. 2-2   An inverter produces an output in the opposite state as the input.

If the high-voltage state represents a one and the low voltage state stands for a zero, the voltage table can be converted to a truth table (see Table 2-3). You can always tell them apart because a voltage table lists the actual voltmeter readings or indicates high and low levels, while a truth table always contains only ones and zeros.

Table 2-3   Inverter Truth Table

Input	Output
0	1
1	0

Conclusions

An inverter complements the input voltage to produce the output. A conducting LED is lighted and drops by about 1.7 V. Voltages for logic gates can be classified as either low or high. Voltage tables list voltmeter readings or high and low levels. Truth tables contain only ones and zeros.

EXPERIMENT 2-2   VISUALIZING AN AND GATE

Purpose

To demonstrate the operation of a diode AND gate.

Parts

Item	Quantity
100-Ω potentiometer	2
LEDs	4
130-Ω resistor	1
Voltmeter	1

Procedure

Step 1.   Wire the circuit shown in Fig. 2-3. The voltmeter is not to be connected at this time. Rotate both potentiometers so that the minimum voltage is applied to points A and B. Both D1 and D2 will glow. What voltmeter readings do you record? You should measure about 0 V.

17

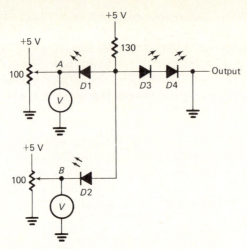

Fig. 2-3 The diode AND gate has a high output if inputs *A* and *B* are both high, so that *D*3 and *D*4 will light; otherwise, the output is low.

Step 2. Are diodes *D*3 and *D*4 on or off? *(Both off)*

Step 3. Rotate the shaft of potentiometer *A* for maximum voltage at that point. What does the voltmeter read? *(Point A = 5 V)* Are diodes *D*1 and *D*2 on or off? *(D1 off, D2 on)* What are the conditions of *D*3 and *D*4? *(Both off)*

Step 4. Reverse the rotation on potentiometer *A* to minimum voltage. Rotate potentiometer *B* for maximum voltage at point *B*. What does the voltmeter measure at each point? *(A = 0 V, B = +5 V)* Record the states of each diode. *(D1 on, D2 off, D3 off, D4 off)*

Step 5. Again reverse the rotation of potentiometer *A*. (Point *A* will now be at maximum voltage.) How do the diodes react? *(D1 off, D2 off, D3 on, D4 on)*

Ins and Outs

The voltage table for the AND gate is given in Table 2-4. As you can read from this table, if both inputs are low, the output is low. Furthermore, if only one input voltage is low, the output level is low. Only if both inputs are high will a high output be produced. The name of the gate comes from the fact that input *A and* input *B* must both be high to produce a high output.

Table 2-4   AND Gate Voltage Table

Input Voltage		Output Voltage
A	*B*	
L	L	L
L	H	L
H	L	L
H	H	H

18

The truth table for the AND gate is derived from the voltage table by substituting zero for low voltage and one for high (see Table 2-5). The truth table conveys the same idea as the voltage table. The symbol for an AND gate (Fig. 2-4) signifies a circuit element that is an implementation of this truth table. Notice that the AND gate can have more than two inputs. In such cases, the truth table is simply extended for the additional inputs. All inputs must be ones before the gate output becomes one, regardless of how many inputs there are.

Table 2-5 AND Gate Truth Table

Input		Output Y
A	B	
0	0	0
0	1	0
1	0	0
1	1	1

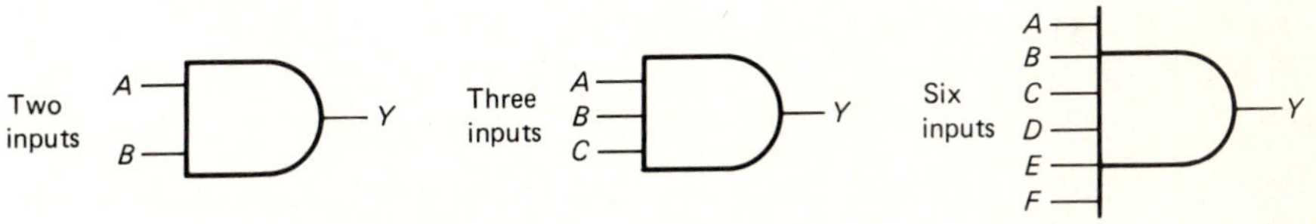

Fig. 2-4   AND gate symbols. No matter how many inputs there are, all must equal one for the output to be one.

Rather than always drawing the symbol for an AND gate, an expression can be written to convey the same relationship between the variables. The AND operation is often shown in three different ways:

$$A \cdot B \qquad (A)(B) \qquad AB$$

In each case the expression is read, "A AND B."

Conclusions

The AND gate truth table shows that the output is zero except when all inputs become one. The operation of the AND gate can be demonstrated with a circuit built of LEDs. The AND relationship is sometimes written in various forms.

EXPERIMENT 2-3   VISUALIZING AN OR GATE

Purpose

To demonstrate the operation of a diode OR gate.

Parts

Item	Quantity
100-Ω potentiometer	2
LEDs	4
130-Ω resistor	1

Procedure

Step 1. Wire the circuit shown in Fig. 2-5. Rotate both potentiometers so that minimum voltage is applied to inputs A and B. Do diodes $D1$ and $D2$ glow? *(No)* What are the conditions of $D3$ and $D4$? *(Both off)*

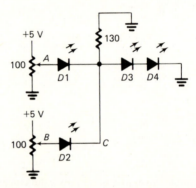

Fig. 2-5 **The diode OR gate produces a high output whenever any input is high. If either the A input *or* the B input is high, $D3$ and $D4$ will light.**

Step 2. Rotate the shaft on potentiometer A to full voltage. What happens to $D1$? *(On)* Are diodes $D3$ and $D4$ on or off? *(Both on)*

Step 3. Reverse the rotation of potentiometer A. Diode $D1$ should go off. Now apply maximum voltage to input B by rotating that potentiometer shaft. What are the states of the diodes $D2$, $D3$, and $D4$? *(All on)*

Step 4. Once more rotate potentiometer A for maximum voltage. What changes do you see in the diodes? *(D1 on)*

Ins and Outs

The voltage table for the OR gate is given in Table 2-6. How does it compare with that of the AND gate? The entries of the OR table can be summarized by saying that the output is high when at least one input is high. Only if all inputs are low does a low output result. The name is derived from the fact that if input A *or* input B is high, the output is high.

Table 2-6   OR Gate Voltage Table

| Input Voltage | | Output Voltage |
A	B	(Point C)
L	L	L
L	H	H
H	L	H
H	H	H

The truth table for this gate can be found by substituting a one for the high level and a zero for the low, just as in the previous cases (see Table 2-7). The symbols for OR gates with varying numbers of inputs are shown in Fig. 2-6. Remember that this symbol represents the truth table and vice versa.

Table 2-7   OR Gate Truth Table

| Input | | Output Y |
A	B	
0	0	0
0	1	1
1	0	1
1	1	1

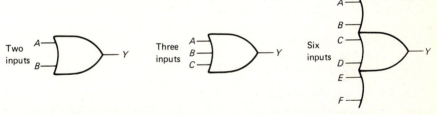

Fig. 2-6   OR gate symbols. Regardless of the number of inputs, the output will be one if any input is one.

The OR function is usually written as $A + B$, read "A OR B." Although the OR sign looks like the plus sign of regular arithmetic, do not let it confuse you. In logic expression, ordinary arithmetic is never used to relate the combinations of inputs.

Conclusions

The OR gate always has an output of one with any input combinations, except when all inputs are zero. The OR sign used in logic expressions looks like the plus sign of ordinary arithmetic, but it does not mean that addition is to be performed.

EXPERIMENT 2-4 USING IC ANDs AND ORs

Purpose

To demonstrate the IC implementation of AND gates and OR gates.

Parts

Item	Quantity
7408 quad AND	1
7432 quad OR	1
SPDT switches	2
Logic probe	1

Procedure

Step 1.   The output levels of these circuits will be tested with a logic probe. You may use the one that you constructed in Chap. 1 or a commercial model. Both the 7408 and the 7432 contain four gates, so they are called *quad AND* and *quad OR* ICs. This experiment uses only one of the gates. Connect circuit 1 as shown in Fig. 2-7*a*. Pay particular attention to hooking pin 7 to a good logic ground and pin 14 to the power supply voltage.

Step 2.   Connect the logic probe to pin 13, and then open and close the switch. What happens? You should see the LED of the logic probe illuminate when the pin is grounded. The LED is off when 5 V is applied to the pin.

Step 3.   Repeat step 2, using pin 12. Note that the LED is on when the input is low and off when the input is high.

Step 4.   Using your probe in the same way on pin 11, complete the voltage table by turning the switches alternately on and off. The output should be low for every input combination except the last one. By comparing Table 2-8 to Table 2-4, can you verify the AND function operation? For any combination of inputs except *A* and *B* high, the AND gate has a low output.

Table 2-8   7408 AND Gate Voltages

Input		Output
A	*B*	
0 V	0 V	*(0 V)*
0	5	*(0)*
5	0	*(0)*
5	5	*(5)*

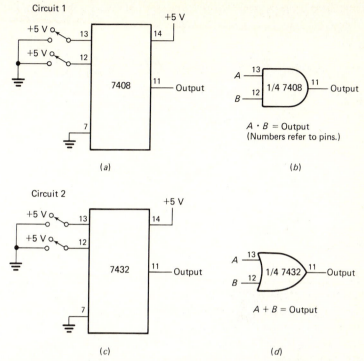

Fig. 2-7 **Integrated circuits which implement OR and AND gates are available. In wiring this experiment, be sure to connect pin 14 to the 5-V power supply and pin 7 to ground.** *(a)* **Schematic;** *(b)* **logic diagram;** *(c)* **schematic;** *(d)* **logic diagram.**

Step 5. Wire the second circuit shown in Fig. 2-7*c*. By again testing the output with the logic probe, complete the voltage table for the 7432 OR gate (see Table 2-9). Match these results against the OR gate in Experiment 2-3. You should find that the output is high whenever one of the inputs is high.

Table 2-9   7432 OR Gate Voltage

Input		Output
A	B	
0 V	0 V	
0	5	
5	0	
5	5	

Ins and Outs

By use of the voltage tables that you obtained, we can construct truth tables by letting a one represent the high level and a zero the low level (see Tables 2-10 and 2-11). By examining these tables, you can discover the general rules for the operation of these gates. These rules are often called *Boolean identities*.

Table 2-10 AND Gate Truth Table

Row	A	B	Output
1	0	0	0
2	0	1	0
3	1	0	0
4	1	1	1

Starting with the AND gate truth table, we see from rows 1 and 3 that

$$x \cdot 0 = 0$$

(where x represents any variable, such as A) because the output will be zero regardless of the value of x. The statement that

$$x \cdot 1 = x$$

is derived from rows 2 and 4. If $x = 0$, then the output is zero; otherwise, x must be one and the output also one.

What about ANDing x with itself? The outcome is

$$x \cdot x = x$$

which is seen in rows 1 and 4 of Table 2-10. Another case involves ANDing x with its complement

$$x \cdot \overline{x} = 0$$

As you found in the expression above, x ANDed with zero is zero. If x is not zero, then $\overline{x}$ must be. Therefore, the output is zero.

Table 2-11 OR Gate Truth Table

Row	A	B	Output
1	0	0	0
2	0	1	1
3	1	0	1
4	1	1	1

Now consider the OR gate truth table. Using the same reasoning as with the AND gate

$$x + 0 = x$$

from rows 1 and 3. Here, however, the result depends on the value of x. What do you think the result of ORing x with one would be? From rows 2 and 4, the answer is

$$x + 1 = 1$$

The solution to ORing x with x should be straightforward for you to find now. From rows 1 and 4

$$x + x = x$$

and a combination of x with its complement yields

$$x + \overline{x} = 1$$

from rows 2 and 3. This result comes about because either x or $\overline{x}$ must be one.

Conclusions

Integrated circuits that implement the AND and OR functions can be obtained. In using these gates, the identities for these two functions can be derived. These identities are summarized in Table 2-12. Figure 2-8 shows how gate circuits can be simplified by using these identities.

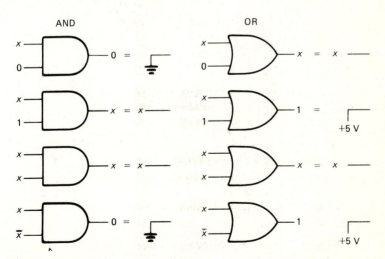

Fig. 2-8 The Boolean identities make it possible to simplify circuits because the output is determined by a single input.

Table 2-12   Boolean Identities

AND	OR
$x \cdot 0 = 0$	$x + 0 = x$
$x \cdot 1 = x$	$x + 1 = 1$
$x \cdot x = x$	$x + x = x$
$x \cdot \overline{x} = 0$	$x + \overline{x} = 1$

EXPERIMENT 2-5   SIMPLIFYING CIRCUITS WITH THE ASSOCIATIVE LAW OF BOOLEAN ALGEBRA

Purpose

To investigate how the order of combination affects logic inputs.

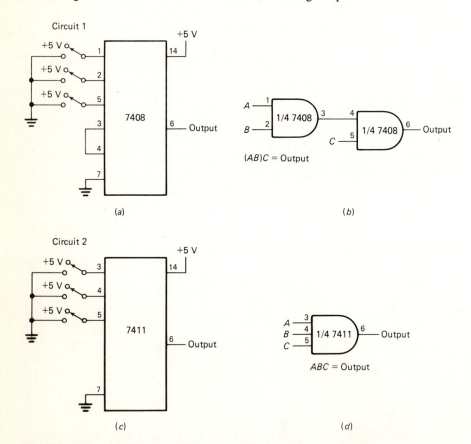

Fig. 2-9   The associative law of Boolean algebra guarantees that the same output is produced by ANDing the inputs in pairs or all at once. (a) Schematic; (b) logic diagram; (c) schematic; (d) logic diagram.

Parts

Item	Quantity
7408 quad AND	1
7411 triple AND	1
SPDT switch	3
Logic probe	1

Procedure

Step 1.   You can use the logic probe that you built in Chap. 1, or a commercially manufactured one, in this experiment. Using the 7408 IC, construct the circuit shown in Fig. 2-9a. Be sure to connect the power supply and ground pins of the 7408, as well as the logic inputs. By closing the switches, you can apply 5 V to the inputs or ground them by opening the switches. The logic probe will show the state of the output switches.

Step 2.   By use of the switches, complete Table 2-13.

Table 2-13   Gate Combination Voltages

	Input		Output
A	*B*	*C*	
0 V	0 V	0 V	
0	0	5	
0	5	0	
0	5	5	
5	0	0	
5	0	5	
5	5	0	
5	5	5	

Step 3.   Now construct circuit 2 shown in Fig. 2-9c and repeat the experiment.

Step 4.   Compare the output columns in Tables 2-13 and 2-14. The output is low except in the last row when *A*, *B*, and *C* are all high. As you can see, the same combination of inputs in each table produces identical output levels.

Ins and Outs

The equivalence of these two circuits is a result of the associative law of Boolean algebra. Using the Boolean notation, the law can be written as

$$(A \cdot B) \cdot C = A \cdot B \cdot C$$

Table 2-14   7411 AND Gate Voltages

	Input		
A	B	C	Output
0 V	0 V	0 V	
0	0	5	
0	5	0	
0	5	5	
5	0	0	
5	0	5	
5	5	0	
5	5	5	

That is, ANDing A with B first and then with C is the same as ANDing them all at once. By extension of the same concept, we can write

$$(A \cdot B) \cdot C = A \cdot (B \cdot C) = A \cdot B \cdot C$$

The law applies equally to OR gates. The circuits to illustrate the effect are similar to those of AND gates. The Boolean notation is

$$(A + B) + C = A + (B + C) = A + B + C$$

Figure 2-10 illustrates these configurations.

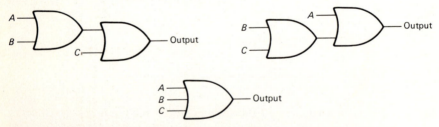

Fig. 2-10   The associate law of Boolean algebra is also valid for OR gates. Any number of inputs may be combined in this manner.

While this relationship may seem to be merely common sense, it does have an important consequence. How many gates were used in circuit 1 as compared with circuit 2? The value of this simple law is that gates can be eliminated in a circuit. By just regrouping the inputs (and using a gate with more inputs), the number of logic elements needed to implement some function can be reduced. The number of inputs that a gate has is called the *fan-in*. A 7408 has a fan-in of 2, while a 7411 has a fan-in of 3.

Conclusions

More efficient circuits can often be developed by using the associative law to group the inputs so as to reduce the number of gates. The law can be used with both AND and OR gates. The number of inputs of a gate is called its fan-in.

EXPERIMENT 2-6 SIMPLIFYING CIRCUITS WITH THE DISTRIBUTIVE LAW OF BOOLEAN ALGEBRA

Purpose

To minimize the number of gates required in a logic circuit constructed of both AND and OR gates.

Parts

Item	Quantity
7408 quad AND	1
7432 quad OR	1
SPDT switch	3
Logic probe	1

Procedure

Step 1. Wire the circuit shown in the schematic in Fig. 2-11*a*. By opening and closing the switches, create each combination of inputs for Table 2-15. Sample the output level with a logic probe.

Table 2-15   Circuit 1 Voltages

	Input		Output
A	B	C	
0 V	0 V	0 V	
0	0	5	
0	5	0	
0	5	5	
5	0	0	
5	0	5	
5	5	0	
5	5	5	

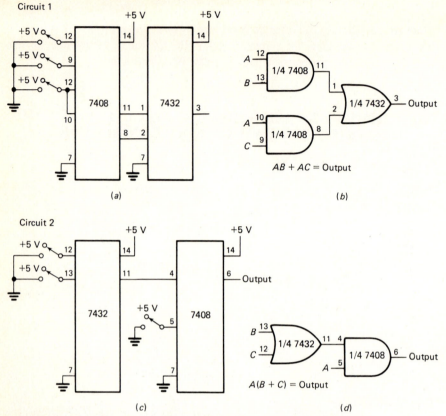

Fig. 2-11 The distributive law of Boolean algebra allows inputs to AND and OR gates to be combined. Often unnecessary circuit components can be removed in this way. (*a*) Schematic; (*b*) logic diagram; (*c*) schematic; (*d*) logic diagram.

Step 2. Using circuit 2 shown in Fig. 2-11*c*, repeat the procedure to fill in the output levels in Table 2-16. Comparing Tables 2-15 and 2-16, notice that the output is high whenever A is high along with either a high input on the B or C terminals.

Ins and Outs

The distributive law of Boolean algebra provides a way of replacing a circuit made up of AND gate and OR gate combinations with a simpler one. By eliminating one of the AND gates, the cost of building the circuit is reduced by one-third (provided that the other gates in the 7408 can be used for other purposes).

The distributive law can be expressed as

$$xy + xz = x(y + z)$$

Written this way, the law relates AND functions that are ORed (see Fig. 2-12). There

30

Table 2-16   Circuit 2 Voltages

	Input		Output
A	B	C	
0 V	0 V	0 V	
0	0	5	
0	5	0	
0	5	5	
5	0	0	
5	0	5	
5	5	0	
5	5	5	

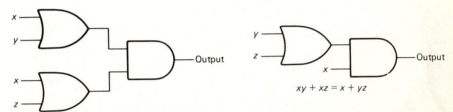

Fig. 2-12   By a combination of functions, the distributive law of Boolean algebra can form an equivalent circuit and cut the number of gates required.

is another form of this law that applies as well to OR gates that are ANDed together. This version may be written as

$$(x + y)(x + z) = x + yz$$

In this case y is ANDed with z first and that result ORed with x to produce the same output conditions as the more complicated statement on the left-hand side of the equality sign. Here, too, the three gates needed for the left-hand expression can be produced by two.

Conclusions

The Boolean algebra laws provide the framework for simplifying combinations of logic elements. The distributive law applies to situations where the output of AND gates are ORed, or in the cases when the outputs of OR gates are ANDed.

EXPERIMENT 2-7   INVESTIGATING INVERTERS

Purpose

To show how IC inverters can complement signals.

Parts

Item	Quantity
7405 inverter	1
4-kΩ resistor	1
SPDT switch	1
Logic probe	1

Procedures

The 7405 contains six inverters, so it is called a *hex inverter* IC. This circuit uses only one of the inverters. Connect the first circuit as shown in the schematic in Fig. 2-13a. Notice that a 4-kΩ resistor is attached between the output and the power supply.

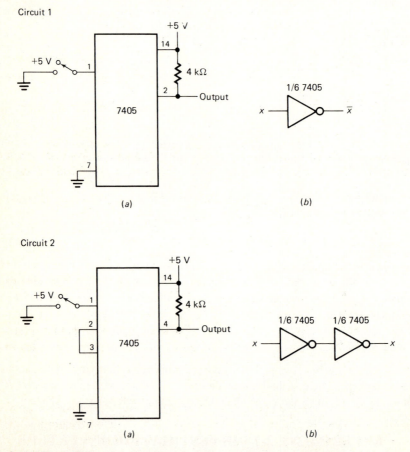

Fig. 2-13 The 7405 is a hex inverter. A pull-up resistor is needed on the output because this is an open-collector circuit. Use of an inverter identity can simplify the second circuit. (*a*) Schematic; (*b*) logic diagram; (*c*) schematic; (*d*) logic diagram.

Step 2. Using the logic probe and the switch, complete the voltage in Table 2-17. The output is the inverse of the input.

Table 2-17   Inverter Voltages

x	Output
0 V	
5	

Step 3. Next, wire circuit 2 shown in Fig. 2-13c. In this configuration, the output of one inverter becomes the input for the next. What readings does the probe give you now (complete Table 2-18)? With a double inversion, the output and the input are identical.

Table 2-18   Combination of Inverters

x	Output
0 V	
5	

Ins and Outs

The inverter chip performs the same function as the discrete component circuit that you built in Experiment 2-1. The truth table for the two circuits is identical (see Table 2-19).

Table 2-19   Inverter Truth Table

x	$\bar{x}$
0	1
1	0

In the second circuit you showed that inverting the input twice produces the same value that you started with. An identity can be written to express the relationship

$$x = \bar{\bar{x}}$$

In fact, any even number of inverter stages (that is, 2, 4, 6, . . .) results in an output equal to the input. An odd number of stages (1, 3, 5, . . .) inverts the input. Figure 2-14 shows how this identity can eliminate redundant inverters in a circuit.

The 7405 gate has a special type of output stage called an *open collector*. An open-collector output stage always requires that a *pull-up* resistor be used in the circuit. More details are provided in Chap. 3, which describes experiments involving output stages of ICs.

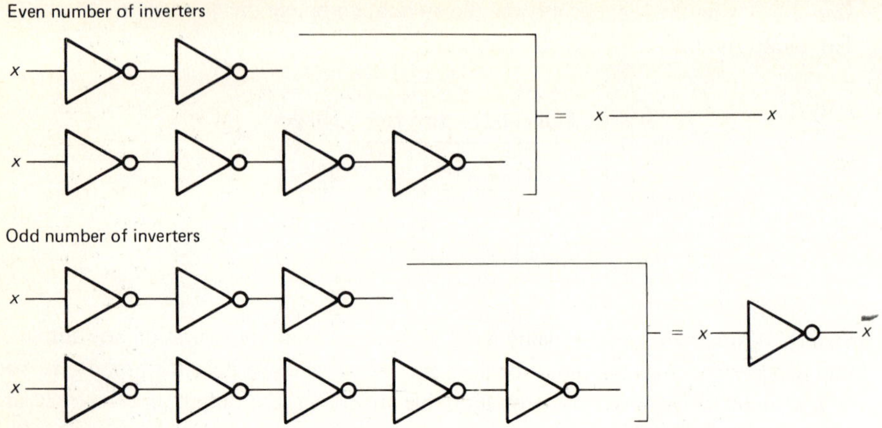

Even number of inverters

Odd number of inverters

Fig. 2-14 The inverter identity makes it possible to remove redundant elements.

Conclusions

The inverter function is also available in the form of an IC. With two inverters, it is possible to demonstrate that inverting the input twice is the same as not inverting it at all. If an IC has an open-collector output, a pull-up resistor must be connected between the output pin and the power supply.

EXPERIMENT 2-8   INVERTING GATE INPUTS

Purpose

To investigate the functions that can be performed by circuits consisting of gates and inverters.

Parts

Item	Quantity
7405 hex inverter	1
7408 quad AND	1
7432 quad OR	1
4-kΩ resistor	2
SPDT switch	2
Logic probe	1

Procedure

Step 1.   Wire circuit 1 shown in Fig. 2-15a. By use of the two switches and the logic probe, fill in the voltages in Table 2-20 for this circuit. You should find that the output is always zero except when A is low and B is high.

Circuit 1

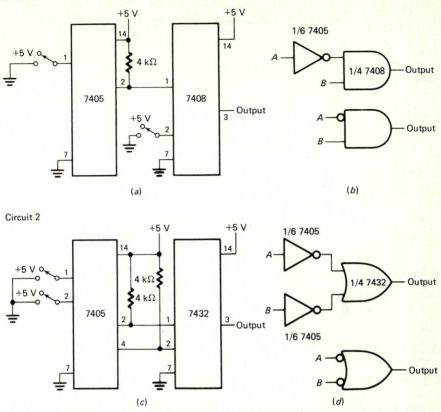

Fig. 2-15 Inverters are frequently used on the inputs of gates. The logic diagrams indicate these inverters with the "bubble" symbol attached to the input terminals. (a) Schematic; (b) logic diagram; (c) schematic; (d) logic diagram.

Table 2-20   Circuit 1 Voltages

A	B	Output
0 V	0 V	
0	5	
5	0	
5	5	

Step 2. Now change the setup to circuit 2 shown in Fig. 2-15b. Repeat the procedure to complete Table 2-21. Here the output is high when either A or B is low.

Table 2-21 Circuit 2 Voltages

A	B	Output
0 V	0 V	
0	5	
5	0	
5	5	

Ins and Outs

The use of inverters on the inputs to a gate changes the truth tables. Figure 2-16 shows all the combinations of inverted inputs that can be used with two-input AND and OR gates. Each symbol is shown with its truth table.

The expressions for these gates show which input is inverted. As examples, the expressions for the gates in Fig. 2-16 can be written:

a. xy e. $x + y$

b. $\bar{x}y$ f. $\bar{x} + y$

c. $x\bar{y}$ g. $x + \bar{y}$

d. $\bar{x}\bar{y}$ h. $\bar{x} + \bar{y}$

The inverted input corresponds to the one with the inverter "bubble."

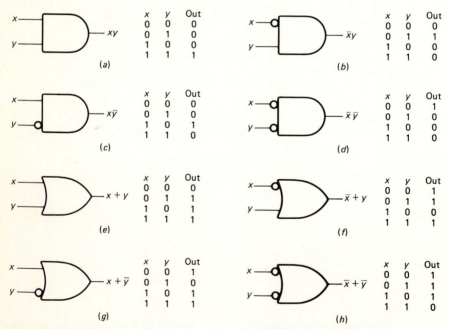

Fig. 2-16   By adding inverters to gate inputs, four different truth tables are developed.

One way to clarify your thinking about these circuits is to realize that any input signal is inverted before applying it to the gate. Consider the circuit in Fig. 2-16b. We can generate that truth table from the ordinary AND gate truth table (see Table 2-22). Each row in the table for the gate with inverted inputs corresponds to a row in the AND gate truth table (see Table 2-23). However, the rows are not in the same order.

Table 2-22

Input		Input After Inversion		Output
x	y	$\overline{x}$	y	
0	0	1	0	0
0	1	1	1	1
1	0	0	0	0
1	1	0	1	0

Generated from AND
truth table

Table 2-23   AND Gate Truth Table

x	y	Output
0	0	0
0	1	0
1	0	0
1	1	1

Conclusions

Inverting the inputs of logic gates produces different truth tables. The inverted input is shown by a bubble on the logic diagram. To find the truth table for the circuit, rewrite the inputs after inversion. Then use the AND or OR truth tables as appropriate.

EXPERIMENT 2-9   USING NAND GATES

Purpose

To investigate the operation of a NAND gate.

Parts

Item	Quantity
7400 quad NAND	1
SPDT switch	2
Logic probe	1

Procedure

Step 1. Connect the circuit shown in Fig. 2-17. Using the switches and the logic probe in the usual manner, complete the voltage in Table 2-24.

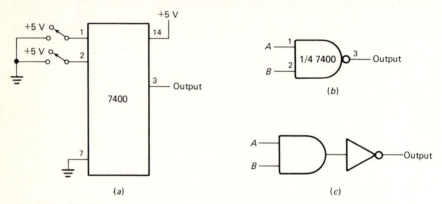

Fig. 2-17 The NAND gate is an AND gate with an inverted output. The same circuit that you can build with two chips is provided in a single IC such as the 7400.

Table 2-24   NAND Voltages

A	B	Output
0 V	0 V	
0	5	
5	0	
5	5	

Step 2. Does your gate have high outputs in every case except when A and B are high?

Ins and Outs

The NAND gate (meaning Not AND) is an inverted AND. The truth tables make this relationship apparent. To find the values in the NAND truth table, simply complement the outputs of an AND gate (see Table 2-25).

The expression for the output of a two-input NAND gate is written $\overline{xy}$. The complement bar covers both inputs. This same convention applies to a four-input NAND which may be written $\overline{wxyz}$. Be sure to notice the difference between $\overline{xy}$ and $\overline{x}\,\overline{y}$.

You can always check relationships between expressions by their truth tables. Table 2-26 (which was also described in the previous experiment) looks nothing like

38

Table 2-25

AND Truth Table			NAND Truth Table		
x	y	Output	x	y	Output
0	0	0	0	0	1
0	1	0	0	1	1
1	0	0	1	0	1
1	1	1	1	1	0

the NAND truth table given in Table 2-25. Be very careful not to confuse these two expressions

$$\bar{x}\,\bar{y} \quad \text{is not equal to} \quad \overline{xy}$$

even though they may *look* very similar.

Table 2-26 $\bar{x}\,\bar{y}$ **Truth Table**

x	y	$\bar{x}$	$\bar{y}$	Output
0	0	1	1	1
0	1	1	0	0
1	0	0	1	0
1	1	0	0	0

Interestingly enough, we have seen the NAND truth table before. Compare the NAND table with that of the circuit in Fig. 2-16h. As you can see, the truth tables are identical, although the logic circuits are quite different. This relationship will be explained in Chap. 3.

Conclusions

Putting an inverter on the output of an AND gate converts it to a NAND function such as the 7400 IC. The expression for NAND is written $\overline{xy}$, which may resemble other AND gate combinations. You can always find out if two logic gate circuits do the same thing by comparing their truth tables.

EXPERIMENT 2-10 USING NOR GATES

Purpose

To investigate the operation of a NOR gate.

Parts

Item	Quantity
7402 quad NOR	1
SPDT switch	2
Logic probe	1

Procedure

Step 1.   Wire the circuit shown in Fig. 2-18. By placing the switches in the correct positions, fill in the voltage in Table 2-27.

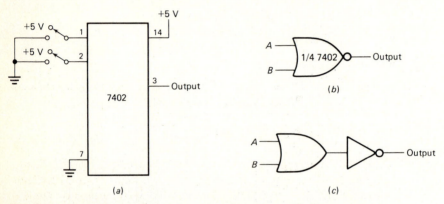

Fig. 2-18   The NOR gate performs the parallel function for OR gates that NAND does for AND gates. The 7402 contains three of these logic elements. *(a)* **Schematic;** *(b)* **logic diagram;** *(c)* **equivalent circuit.**

Table 2-27   OR Gate Voltages

A	B	Output
0 V	0 V	
0	5	
5	0	
5	5	

Step 2.   Did you find the output low except when the *A* and *B* inputs are 0 V?

Ins and Outs

The NOR function is written as

Two inputs	Four inputs
$\overline{x + y}$	$\overline{w + x + y + z}$

40

Here, too, the complement bar covers all the inputs. Can you write the truth table for a NOR gate (see Table 2-28)? The table is just an OR function with the outputs complemented. (This is a parallel to the NAND function of the last experiments.) In the equivalent circuit (Fig. 2-18c), the inverter on the output can be explicitly seen.

Table 2-28

OR Gate Truth Table			NOR Gate Truth Table		
x	y	$x + y$	x	y	$\overline{x + y}$
0	0	0	0	0	1
0	1	1	0	1	0
1	0	1	1	0	0
1	1	1	1	1	0

Just as with the NAND, other expressions may look much the same as the NOR but not be equal:

$$\overline{x} + \overline{y} \quad \text{is not equal to} \quad \overline{x + y}$$

You can prove this inequality by comparing the NOR truth table in Table 2-28 with that in Fig. 2-16h. Note that they are not the same. Does any other table in that figure correspond to the NOR? Yes. The AND gate with both inputs inverted is identical. The reason for this equivalence will also be explained in Chap. 3.

Conclusions

The NOR gate, such as a 7402, is an OR gate with an inverted output. Now that you have completed these experiments, the basic logic functions of NOT, AND, OR, NAND, and NOR should be meaningful. In Chap. 3, you will be experimenting further with the TTL versions of these gates.

3

5400/7400 Integrated Circuits

The 5400/7400 family of ICs is a compatible series of TTL circuits that implement a wide variety of logic functions. The designation 5400/7400 refers to the numbering scheme used for these components. The number can be divided into two parts, one indicating temperature range and the other the function performed. For example,

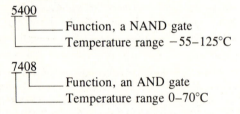

5400
└─┴── Function, a NAND gate
└──────── Temperature range −55–125°C

7408
└─┴── Function, an AND gate
└──────── Temperature range 0–70°C

As you can see, the 5400 series can tolerate wider temperature extremes than can the 7400 series. Other than this characteristic, two chips with the same function identifier, such as 7408 and 5408, can be used interchangeably. As you would expect, though, the 7408 is less expensive.

In this chapter you will exercise these gates in a series of experiments that illustrate the electrical and logic features of the IC family. Concepts such as propagation delay, wired logic, and equivalent circuits are examined. Then the ways in which binary quantities can be compared to find their relative size are demonstrated. A detailed look at three-state and strobed logic is also considered in these investigations.

EXPERIMENT 3-1   VERIFYING DE MORGAN'S THEOREM

Purpose

To investigate equivalent circuits by means of De Morgan's theorem.

Parts

Item	Quantity
7410 triple NAND	1
SPDT switch	3
Logic probe	1

Procedure

Step 1. Construct the circuit shown in Fig. 3-1a. The switches permit you to change the inputs to the 7410 gate. With the aid of a logic probe, complete the voltage table for this circuit (see Table 3-1).

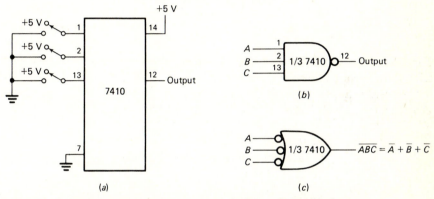

(a)

(b)

(c)

Fig. 3-1 De Morgan's theorem states an equivalent relationship between NAND gates and OR gates with inverted inputs. The theorem also applies to NOR gates which have truth tables identical to those of AND gates with inverted inputs. *(a)* Schematic; *(b)* logic diagram; *(c)* equivalent circuit.

Table 3-1   7410 Gate Voltages

	Input		Output
A	B	C	
0 V	0 V	0 V	
0	0	5	
0	5	0	
0	5	5	
5	0	0	
5	0	5	
5	5	0	
5	5	5	

Step 2. Did you find that outputs are high except when every input is 5 V? From your background of Chap. 2, you should be able to construct the NAND gate truth table from the voltages (see Table 3-2).

Table 3-2 NAND Gate Truth Table

	Input		
A	B	C	Output
0	0	0	(1)
0	0	1	(1)
0	1	0	(1)
0	1	1	(1)
1	0	0	(1)
1	0	1	(1)
1	1	0	(1)
1	1	1	(0)

Step 3. Next find the truth table for the logic diagram in Fig. 3-1c (see Table 3-3). Compare it to Table 3-2. Did you find that they are identical?

Table 3-3 Comparison Truth Table

A	B	C	Output
0	0	0	
0	0	1	
0	1	0	
0	1	1	
1	0	0	
1	0	1	
1	1	0	
1	1	1	

Ins and Outs

You may think that an equivalence between NAND gates and OR gates is unexpected. The rule that governs the correspondence of these two truth tables and, therefore, the two gates, is called *De Morgan's theorem*. This rule says that you can consider these two circuits identical because

$$\overline{ABC} = \overline{A} + \overline{B} + \overline{C}$$

44

The terms need not be limited to three inputs as in this case. In general,

$$\overline{ABCD \ldots} = \overline{A} + \overline{B} + \overline{C} + \overline{D} + \ldots$$

where the dots mean "continued as far as you like."

An easy way to remember the theorem is to observe that all ANDs on the left-hand side of the equality sign become ORs on the right. Also, the long complement bar is broken into separate bars over each term.

The theorem also applies in a similar fashion to NOR gates. The rule here is

$$\overline{A + B + C + D + \ldots} = \overline{A}\,\overline{B}\,\overline{C}\,\overline{D} \ldots$$

That is, change ORs to ANDs and again break the large complement bar into individual ones.

These rules can be applied to mixed expressions as well. For example,

$$\overline{(X + Y) \cdot Z} = \overline{X}\,\overline{Y} + \overline{Z}$$

$$\overline{\overline{fe} + h} = (\overline{\overline{f}} + \overline{e}) \cdot \overline{h} = (f + \overline{e})\,\overline{h}$$

$$\overline{\overline{km} + np} = \overline{(\overline{k} + \overline{m})} + np = \overline{\overline{k}}\,\overline{\overline{m}}\,(\overline{n} + \overline{p}) = km\,(\overline{n} + \overline{p})$$

In the second case, the $\overline{f}$ term gets complemented again, so in the final expression just f appears. The last example shows how two complement bars are handled. First the inner bar is broken giving $\overline{k} + \overline{m}$. Then the outer bar is processed, with care taken to group the terms correctly. The double complement bars over k and m are also eliminated.

"Well," you may ask, "if the two symbols in the logic diagrams are both right, is this a NAND gate or an OR gate?" Usually the symbol for the NAND gate is shown. This classification is also the one used by the manufacturer, so showing a 7410 as a NAND in logic diagrams is the most straightforward. For special purposes, though, the logic diagram may occasionally indicate a 7410 as an OR gate with inverted inputs.

Conclusions

De Morgan's theorem permits long complement bars to be divided into the complement of each individual term. Furthermore, each OR becomes an AND, and every AND becomes an OR in the equivalent expression. Most often logic diagrams are drawn with these gates as NANDs and NORs, but sometimes the equivalent circuits are indicated in their place.

EXPERIMENT 3-2   OUTPUT STAGES

Purpose

To examine the two types of outputs used with TTL.

Parts

Item	Quantity
7408 quad AND	1
7409 quad AND	1
SPDT switch	2
130-Ω resistor	1
4-kΩ resistor	1
LED	1

Procedure

Step 1. Connect circuit 1 as shown in Fig. 3-2*a*. The LED indicates the output level. It glows when the output is high and is off in response to a low output. With the switches in various combinations and from noting the LED condition, complete the voltages in Table 3-4.

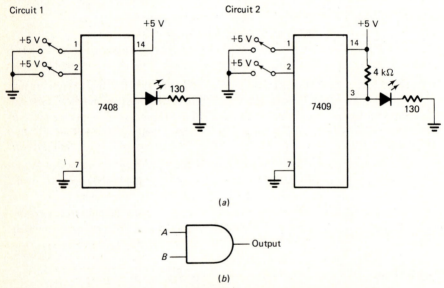

(a)

(b)

Fig. 3-2 The 7408 AND has totem-pole output stages, while the 7409 has open-collector output stages. *(a)* Schematic; *(b)* logic diagram.

Table 3-4   7408 Voltages

Input		Output
A	*B*	
0 V	0 V	*(0 V)*
0	5	*(0)*
5	0	*(0)*
5	5	*(5)*

Did your table (Table 3-4) agree with the expected one for an AND gate?

Step 2. Now replace the 7408 IC with the 7409. Repeat the first step. Do *not* put the 4-kΩ resistor in the circuit yet. What are your results? You should find that the output is always low.

Step 3. Put the 4-kΩ resistor in place. Your results should now agree with those in Table 3-4. Why do you think that the name ''pull-up'' is given to the resistor?

Ins and Outs

The only difference in these two ICs is the final transistor stage before the output pin. These are shown in Fig. 3-3, but the rest of the gate circuit is merely outlined by a box.

The totem-pole output operates in a push-pull mode. When Q1 and Q2 are saturated, Q3 is cut off. Remember that the voltage across the collector-emitter of a saturated transistor is only a few tenths of a volt, so Q1 pulls the output to almost ground potential, while Q3 does not allow any current to flow. When Q3 saturates, Q1 and Q2 are cut off. The states of the transistors isolate the output voltage from ground, so no current flows. Saturated transistor Q3 causes a high output. (With no current flow, there is no voltage dropped across the 100-Ω resistor.)

The open collector functions in the same way when Q1 and Q2 are saturated and the output is pulled to ground. When Q1 and Q2 are off, there is no output without a pull-up resistor. The ''strange'' results you found in Step 2 demonstrated that there is no voltage at the output terminal when the pull-up resistor is omitted. When in place, the resistor provides a path for the power supply current.

A major difference between these two ICs is their speed of operation. Typical switching time from low to high for the 7408 is 17.5 nanoseconds (ns), while the 7409

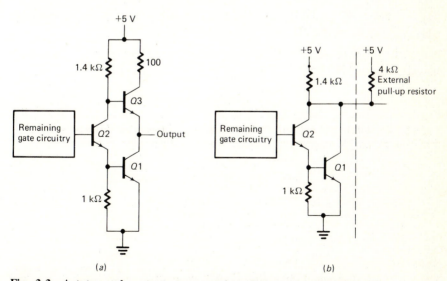

(a)　　　　　　　　　　　　　　(b)

Fig. 3-3 A totem-pole output stage requires that more transistors and resistors be built into the chip. An external resistor is needed with the simpler open-collector output. Surprisingly, both ICs usually cost the same amount. (a) Totem-pole output (simplified); (b) open-collector output.

requires 21 ns. The difference is a 20 percent speed increase with the totem-pole output. Why, then, are open collectors used? The next experiment will help you answer that question.

Conclusions

Totem-pole output stages do not require external pull-up resistors, while open-collector output stages do. The use of a larger output resistor and absence of push-pull action makes the open collector the slower of the two.

EXPERIMENT 3-3   WIRED LOGIC

Purpose

To investigate wired AND logic.

Parts

Item	Quantity
7409 quad AND	1
4-kΩ resistor	1
SPDT switch	4
Logic probe	1

Procedure

Step 1.   From your knowledge of AND gates, fill in the output to complete the truth table (see Table 3-5) for Fig. 3-4.

Step 2.   Now wire the circuit shown in the schematic of Fig. 3-5a. How do its outputs compare with your truth table? They should be identical.

CAUTION

Do not use the 7408 totem-pole output IC in this experiment.

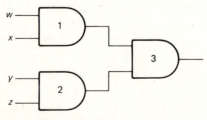

Fig. 3-4   Logic diagram of three AND gates.

Table 3-5   Three AND Gate Truth Table

Gate 1		Gate 2		
w	x	y	z	Output
0	0	0	0	(0)
0	0	0	1	(0)
0	0	1	0	(0)
0	0	1	1	(0)
0	1	0	0	(0)
0	1	0	1	(0)
0	1	1	0	(0)
0	1	1	1	(0)
1	0	0	0	(0)
1	0	0	1	(0)
1	0	1	0	(0)
1	0	1	1	(0)
1	1	0	0	(0)
1	1	0	1	(0)
1	1	1	0	(0)
1	1	1	1	(1)

By simply tying the outputs of gates 1 and 2 together, the effect of another AND gate is produced. To understand how this unexpected combination happens, refer to the diagram for an open-collector output stage. When output transistor $Q1$ saturates, it pulls the output of all gates tied to a common point low. That transistor sinks the current from the power supply of any gate that is on; in this example, gate 1 is pulled low (see Fig. 3-5c).

In using wired logic you must be careful to not exceed the current handling capacity of $Q1$. As more and more gates are tied into a wired AND, all except one could be high. The output transistor in the remaining gate must sink all the current. From Ohm's law, the approximate output current of a high gate can be found

$$ I = \frac{5 \text{ V}}{4 \text{ k}\Omega} = 1.25 \text{ mA} $$

Notice that this current depends on the value of the pull-up resistors. Normally the output transistor can be expected to sink around 11 mA of current. (Always check the

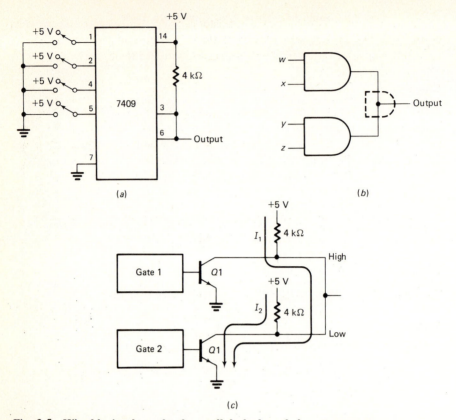

Fig. 3-5 Wired logic, shown by the small dashed symbol, can save gates.
There are disadvantages in using this technique, too. (a) Schematic; (b) logic
diagram; (c) open-collector output stage.

manufacturer's data sheet for any circuit you design yourself.) Then the number of
AND gates that can be wired together is

$$\frac{11 \text{ mA total capacity} - 1 \text{ mA}}{1.25 \text{ mA for one gate}} = 8 \text{ gates}$$

The 1 mA must be subtracted from the numerator because the next downstream gate
can also supply current to the sink.

The chief shortcoming of wired logic is that there is no way to find a defective gate
from a group wire-ANDed together without disconnecting them. After a few disas-
semblies, some gates are damaged from the heat of the soldering iron. Other problems
associated with wired logic are increased noise and slower switching speeds.

An important point to note is that totem-pole output stages can *never* be used in
wired logic configurations. Because the resistor in the output stage is only on the
order of just 100 Ω, the output-stage transistor of the low gate would have to sink
some 50 mA:

$$\frac{5 \text{ V}}{100} = 50 \text{ mA}$$

50

This current greatly exceeds what the transistor can safely sink, and the IC would be destroyed if the current were allowed to flow. Therefore, a primary purpose of open-collector gates is to allow the designer to use wired logic.

Conclusions

Wired logic makes it possible to reduce the number of gates by wiring the outputs to a common point. The wire AND formed is identified with a dashed symbol on the schematic. Only open-collector TTL gates can safely be used to implement wired logic functions; gates with totem-pole outputs should never be used in this way.

EXPERIMENT 3-4 MEASURING THE TIME DELAY IN A TTL GATE

Purpose

To show that signals do not travel through logic gates instantaneously.

Parts

Item	Quantity
7400 quad NAND	1
Oscilloscope	1

Procedure

Step 1. Wire the circuit shown in Fig. 3-6a. The output goes to the vertical input of the scope and is also fed back to pin 1. Set the vertical sensitivity for 0.5- to 1-V division.

Step 2. Adjust the time base so that the pulse train is clearly shown.

Step 3. Measure the period of the waveform. If your scope is not fast enough to measure the waveforms in this experiment, additional gates can be cascaded. An odd number of gates is required for the proper output.

Step 4. Convert this period to the propagation delay in one gate:

$$\text{Delay} = \frac{\text{period of waveform}}{2 \times \text{number of gates}}$$

For example, if you have three gates in the circuit and measure a period of 66 ns, then

$$\text{Delay} = \frac{66}{2 \times 3} = 11 \text{ ns}$$

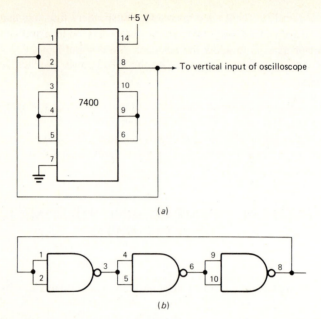

+5 V

7400

To vertical input of oscilloscope

(a)

(b)

Fig. 3-6 The three NAND gates in this circuit are connected to act as inverters. A continual train of output pulses is produced. (a) Schematic; (b) logic diagram.

Ins and Outs

The propagation delay time is the interval between specified reference points on the input and output voltage waveforms. Usually the time in changing from low to high level is not equal to that for a high-to-low transition.

With standard TTL gates, the propagation delay is on the order of 10 ns. There are other series of compatible TTL circuits with faster and slower propagation times. Table 3-6 compares the typical performance characteristics for 54/74 family gates.

Table 3-6 54/74 Family Performance Characteristics

Series	Designator	Propagation Time (ns)	Power Dissipation [milliwatts (mW)]
Standard	54/74	10	10
High-speed	54H/74H	6	22
Low-power	54L/74L	33	1
Schottky	54S/74S	3	19
Low-power Schottky	54LS/74LS	9.5	2

As you can see in Table 3-6, increasing the speed of the gate also requires an increase in power. The low-power Schottky series has the best performance in both speed and power.

Integrated circuits can be compared by the speed-power product. In the case of standard TTL,

$$\text{Speed-power product} = 10 \text{ ns} \times 10 \text{ mW} = 100 \text{ picojoules (pJ)}$$

For low-power Schottky, the product is

$$9.5 \text{ ns} \times 2 \text{ mW} = 19 \text{ pJ}$$

The tradeoff between speed and power for the TTL families is summarized in the graphs in Fig. 3-7.

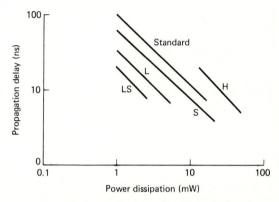

Fig. 3-7 **Increasing speed for a TTL gate means that the power consumption also goes up. The graph shows how the power and speed for one gate can vary for different TTL families.**

Conclusions

The average propagation delay of logic gates can be measured by using a circuit like the one in this experiment. The delay in standard TTL is on the order of 10 ns. The high-speed Schottky and low-power Schottky series are faster, while the low-power series is slower. The designator of the IC identifies the series it is in with one or two letters such as H, L, or LS.

EXPERIMENT 3-5   MAKING AN EXCLUSIVE OR

Purpose

To become familiar with combining simple logic functions to form more powerful operations.

Parts

Item	Quantity
7400 quad NAND	1
SPDT switches	2
Logic probe	1

Procedure

Step 1. Wire the circuit shown in Fig. 3-8a.

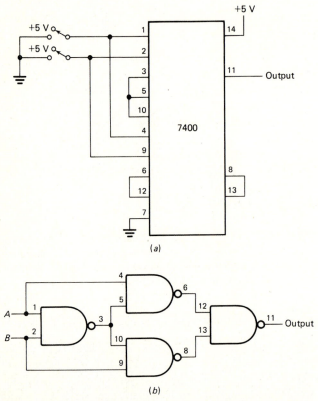

(a)

(b)

Fig. 3-8 One form of the exclusive OR is constructed by interconnecting four NAND gates. The output of the exclusive OR is one whenever the A and B inputs are not equal. (a) Schematic; (b) logic diagram.

Step 2. Test your circuit by opening and closing the switches in all combinations to complete the truth table (see Table 3-7).

Step 3. How does Table 3-7 differ from that of the OR gate you investigated earlier? You should find that the only difference is the output for the last row. Both

Table 3-7 Exclusive OR Truth Table

Input		Output
A	*B*	
0	0	*(0)*
0	1	*(1)*
1	0	*(1)*
1	1	*(0)*

inputs equal to one produce a one output of an OR gate, but they result in a zero with the exclusive OR.

Ins and Outs

The exclusive OR is a widely used circuit. Because it appears so frequently in diagrams, a special symbol has been defined for it (see Fig. 3-9). The expression for the circuit in Fig. 3-8 may be written

$$\text{Output} = \bar{A}B + A\bar{B}$$

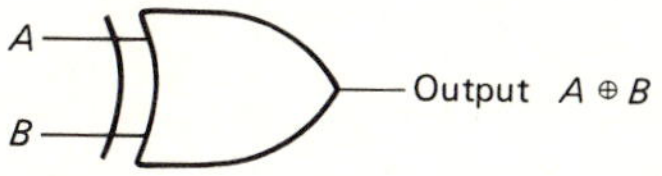

Fig. 3-9 Exclusive OR symbol.

Often the relationship is shown with the exclusive OR sign instead:

$$\text{Output} = A \oplus B$$

Conclusions

The exclusive OR differs from the OR gate in that the output is one only when both inputs are not equal. The frequent appearance of the exclusive OR in schematics has brought a special symbol for the function into use. Exclusive OR expressions can be written with the $\oplus$ symbol. The exclusive OR makes it possible to test whether the inputs are equal or not. For that reason, this function can be thought of as an inequality testing device.

EXPERIMENT 3-6 SHOWING POSITIVE AND NEGATIVE LOGIC EFFECTS

Purpose

To learn how to use truth tables for positive or negative logic.

Parts

Item	Quantity
7410 triple NAND	1
7405 hex inverter	1
SPDT switch	3
Logic probe	1

Procedure

Step 1. Wire the circuit as shown in Fig. 3-10*a*. As you have done before, fill in the voltage table for the circuit (see Table 3-8).

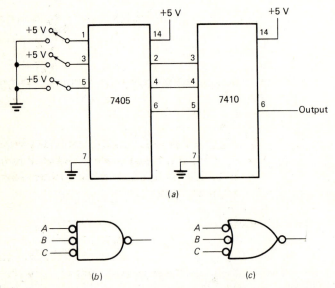

Fig. 3-10   Changing between positive and negative logic converts functions to a different form. (*a*) Schematic; (*b*) positive logic diagram; (*c*) negative logic diagram.

Step 2. Convert the voltage table to a truth table by letting 0 V stand for zero and 5 V represent one (see Table 3-9).

Did your table agree with the expected for a three-input NAND gate with inverters on each input?

Step 3. Build another truth table (see Table 3-10), but this time let the 0-V entry in the voltage table be represented by a one in the truth table. The 5-V values then must be changed to zeros. The first row of the table is done for you.

Table 3-8   Voltage Table

	Input		Output
A	B	C	
0 V	0 V	0 V	(0)
0	0	5	(5)
0	5	0	(5)
0	5	5	(5)
5	0	0	(5)
5	0	5	(5)
5	5	0	(5)
5	5	5	(5)

Table 3-9   Positive Logic Truth Table

	Input		Output
A	B	C	
0	0	0	(0)
0	0	1	(1)
0	1	0	(1)
0	1	1	(1)
1	0	0	(1)
1	0	1	(1)
1	1	0	(1)
1	1	1	(1)

Table 3-10   Negative Logic Truth Table

	Input		Output
A	B	C	
1	1	1	1
1	1	0	(0)
1	0	1	(0)
1	0	0	(0)
0	1	1	(0)
0	1	0	(0)
0	0	1	(0)
0	0	0	(0)

Ins and Outs

Table 3-9 represents the positive logic form of the gate under test. To convert a voltage table to positive logic, let the more positive voltage be represented as a one. Table 3-10 is a negative logic truth table. In this case, the more negative logic represents a one.

What function does Table 3-10 perform? It is the three-input NOR gate with inverted inputs shown in Figure 3-10c?

In general, changing the notation from positive to negative causes each AND to become an OR and each OR to become an AND. The same thing happens in going from negative to positive logic. Figure 3-11 summarizes all possible two-input combinations in this way. You may want to check some of these results by drawing up the voltage and truth tables.

So far in this book, positive logic has been used. Most manufacturers also specify their gates by use of positive logic. For that reason the 7410 is a *NAND gate*; however, it could also correctly be called a *negative logic NOR gate*. So, which is right? Neither one is more right than the other. This seeming contradiction is merely

Conversion between positive and negative logic

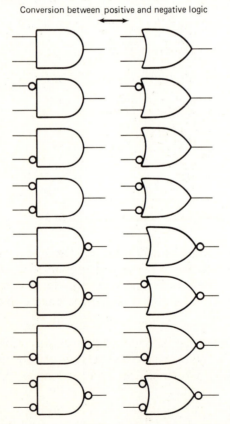

Fig. 3-11   Conversions between positive and negative logic interchanges OR functions with ANDs.

an artifact of the convention used (much like the difference between metric and U.S. Customary units). Just so you are consistent, either method of analysis produces the correct answer regarding gate output. Positive logic is used most frequently, so this book will stay with positive logic most of the time. At times negative logic can simplify the circuit action. To keep everything clear, any exceptions to the use of positive logic will be indicated.

Conclusions

Positive and negative logic are just ways of assigning binary values to voltages. If the more positive voltage is considered to be a one in the truth table, positive logic is being employed. The effect of converting from positive to negative logic is to cause OR functions to be transformed to AND gates and AND gates to OR gates. This seeming alteration is just the result of a convention. The electronic circuit and its voltage table are in no way modified, regardless of the type of logic used in the analysis.

EXPERIMENT 3-7   TESTING THREE-STATE LOGIC

Purpose

To become familiar with three-state logic devices.

Parts

Item	Quantity
74126 three-state quad buffer	1
7410 triple NAND	1
SPDT switches	4
Logic probe	1
Voltmeter	1

Procedure

Step 1.   Wire the circuit shown in the schematic of Fig. 3-12a. With the strobe (Pins 1, 4, and 9) connected to the 5-V supply, measure the output for every combination of the A, B, and C inputs. Use this information to complete the voltages in Table 3-11. Also, what is the voltage from any one of the buffers when its input is high? When its input is low? (*High, Low*)

Step 2.   Repeat the experiment, but this time ground the strobe before changing the inputs. Did you find that the output did not change from high to low? What are the voltages from any buffer? (*No change for either high or low input*)

Ins and Outs

Three-state TTL (also called *Tristate* TTL) can effectively disconnect itself from the remainder of the circuit. In addition to the high- and low-output situations that you

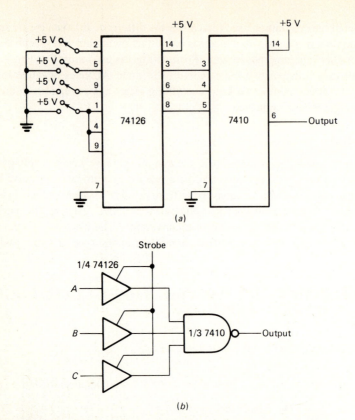

Fig. 3-12 **Three-state gates with normal outputs when the strobe is high. When the strobe is low, the output is disabled and the gate is in the high impedance state.** *(a)* Schematic; *(b)* logic diagram.

Table 3-11   Voltage Table

	Input		
A	B	C	Output
0 V	0 V	0 V	(5 V)
0	0	5	(5)
0	5	0	(5)
0	5	5	(5)
5	0	0	(5)
5	0	5	(5)
5	5	0	(5)
5	5	5	(0)

have already seen, this series of gates has a high-impedance state. In this latter condition almost no current can flow in or out [typically less than 50 microamperes (μA)], so the gate is isolated.

The most frequent use of three-state logic is in bused systems. A simple example of such a system is illustrated in Fig. 3-13. Because the outputs of buffers 2, 3, and 4 are disabled, no outputs or inputs are permitted gates B and C. However, the buffers associated with gates A and D are enabled, so normal logic functions are performed. At some other time, buffer 1 could be disabled and 2 enabled. Then gate B could send to D. By such combinations of strobe inputs, any one sending gate can transmit to only one of the receiving gates. Control of the bus in this manner is essential to prevent confused signaling, such as would result if several gates attempt to transmit on the bus at one time.

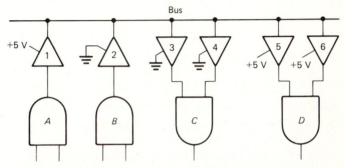

Fig. 3-13   A simple bused system. Only gate A is permitted to send data to gate D during the period shown.

Conclusions

Three-state logic provides a high-impedance state that disables gate inputs or outputs. Three-state devices are frequently encountered in bused circuits.

EXPERIMENT 3-8   DESIGNING AN EQUALITY CIRCUIT

Purpose

To build a circuit that tests the equality of binary inputs.

Parts

Item	Quantity
7408 quad AND	1
7405 hex inverter	1
7432 quad OR	1
SPDT switch	2
Logic probe	1

Procedure

Step 1. In this experiment, the procedure will be the reverse of the previous ones. Instead of being given a circuit to test, you will develop the requirements first. Then they will be verified. The desired equality circuit should have a positive truth table such as Table 3-12.

Table 3-12   Equality Truth Table

Inputs		Output
A	B	
0	0	1
0	1	0
1	0	0
1	1	1

That is, the output is one whenever the two inputs are the same.

Step 2. When this sort of problem is given, one way to proceed is to observe that only the first and last rows are of interest. If the output for only these two conditions (both inputs zero or one) could be guaranteed, the other two rows would also be satisfied. As you will notice, the output is to be one when both inputs are in the same state. This requirement can be written

$$\overline{A}\,\overline{B} + AB$$

(If the input is zero in the truth table, that term is complemented in the expression.) To check the expression, find the output if A and B are the same. Regardless whether the inputs are zero or one, the output is one.

$$\overline{0} \cdot \overline{0} + 0 \cdot 0 = 1$$
$$\overline{1} \cdot \overline{1} + 1 \cdot 1 = 1$$

Step 3. Convert the expression to a logic diagram.

Step 4. Look up the IC diagrams in Appendix A to find the pin numbers for gate inputs and outputs. One such solution is shown in Fig. 3-14b. Other AND gate inverters and OR gates in the IC packages could have been used as well.

Step 5. Change the truth table to a voltage table. Verify that operation is satisfactory.

Ins and Outs

The equality circuit is the complement of an exclusive OR. You can confirm this claim by comparing the two truth tables. The equality circuit often appears in logic diagrams as the symbol shown in Fig. 3-14c. The relationship between the variables is written

$$A \odot B$$

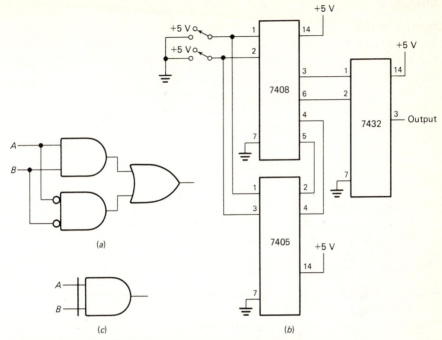

Fig. 3-14 This circuit is one way to realize an equality function. Once the logic diagram is drawn, a simple check of IC diagrams gives the pin numbers. (*a*) Logic diagram; (*b*) schematic; (*c*) circuit symbol.

Conclusions

To design a circuit which satisfies specified conditions, you begin with the truth table. Working only with the rows that have ones as outputs, write the expression for all the input combinations which produce that output. Finally choose ICs which perform the necessary functions. The equality function tests whether all its inputs are equal or not. Unique symbology can be used to indicate this function, as Fig. 3-14*c* shows.

EXPERIMENT 3-9   CHECKING DIGITAL COMPARATOR OPERATION

Purpose

To make a series of digital comparison circuits.

Parts

Item	Quantity
7408 quad AND	1
7405 hex inverter	1
SPDT switch	2
Logic probe	1

Procedure

Step 1. Wire the circuit shown in the schematic of Fig. 3-15. Find the voltage table for the circuit.

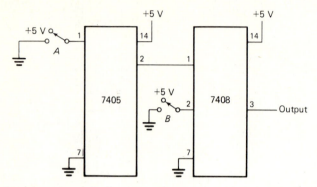

Fig. 3-15   One of a series of digital comparison circuits called *comparators*.

Step 2. What is the positive logic truth table for the circuit (see Table 3-13)?

Table 3-13   Truth Table

Input		Output
A	*B*	
0	0	*(0)*
0	1	*(1)*
1	0	*(0)*
1	1	*(0)*

Step 3. What is the relative size of the inputs when the output is equal to one? *(B is greater than A)*

Ins and Outs

Did you find that *A* is zero and *B* is one when the output level is one? By so doing, you demonstrated that the circuit tests whether *B* is greater than *A*. Then the expression

$$\overline{A}B \quad \text{is equal to} \quad B > A$$

By moving the inverter to the other input, you can test for $A > B$ as shown in Fig. 3-16. If an OR gate is used, you can also make a test for greater than or equal to. You have also already seen how the equality and exclusive OR can measure equality or nonequality of inputs. You may wish to build some of these other circuits to verify the truth tables.

Can you figure out how to determine if a particular signal is equal to one (in positive logic)? One way would be to tie the *A* input of the $\overline{A} + B$ circuit to 5 V and apply the

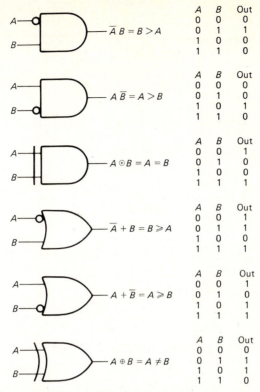

A	B	Out
0	0	0
0	1	1
1	0	0
1	1	0

$\overline{A}\,B = B > A$

A	B	Out
0	0	0
0	1	0
1	0	1
1	1	0

$A\,\overline{B} = A > B$

A	B	Out
0	0	1
0	1	0
1	0	0
1	1	1

$A \odot B = A = B$

A	B	Out
0	0	1
0	1	1
1	0	0
1	1	1

$\overline{A} + B = B \geqslant A$

A	B	Out
0	0	1
0	1	0
1	0	1
1	1	1

$A + \overline{B} = A \geqslant B$

A	B	Out
0	0	0
0	1	1
1	0	1
1	1	0

$A \oplus B = A \neq B$

Fig. 3-16 A complete set of two-input digital comparators.

unknown signal to the *B* input. Of course, this is a somewhat indirect method, because the signal could also be directly tested for a high level.

Conclusions

Every possible measure of the relative values of binary quantities can be tested with the comparators. Comparators are common computer and microprocessor components.

EXPERIMENT 3-10   CONTROLLING GATE TIMING

Purpose

To get experience with strobe gates.

Parts

Item	Quantity
7423 quad NOR	1
SPDT switch	5
Logic probe	1

Procedure

Step 1. Connect the circuit in the diagram for circuit 1 in Fig. 3-17*a*. Connect the switch on pin 12 to 5 V. Test the output using various inputs.

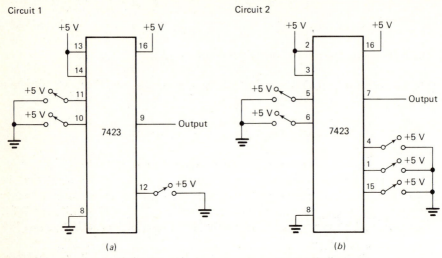

Fig. 3-17 **Strobed gates develop their outputs only when the strobe conditions are true.**

Step 2. Now ground the switch on pin 12.

Step 3. Again test the output for various input combinations. Do you find that the level is always high?

Step 4. Next wire the test configuration to agree with circuit 2 in Fig. 3-17*b*. Connect the switches on pins 4 and 15 to 5 V, the one on pin 1 to ground.

Step 5. Again test for the NOR gate function. You should discover normal operation.

Step 6. One by one, reverse the switch settings on pins 1, 4, and 15. After each switch is changed, attempt to verify the previous result. (Before testing the outcome of the next switch reversal, return the previous one to the position of Step 4.)

Ins and Outs

The 7423 is a dual expandable NOR gate with a strobe (see Fig. 3-18). It is intended to be used with a 7460 dual four-input expander. The expander permits a larger fan-in than do the four inputs of the 7423.

The NOR gate you tested first has a strobe on pin 12. Only when the strobe is high will the output change. The expression for the output is

$$\text{Output} = \overline{\text{strobe} \cdot (A + B + C + D)}$$

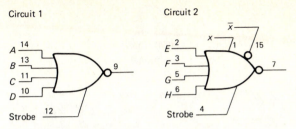

Fig. 3-18 Logic diagrams of the two gates in a 7423 IC.

The gate in circuit 2 has even more control inputs. As shown in Fig. 3-18, its output depends on

$$\text{Output} = \overline{\text{strobe} \cdot (E + F + G + H) + x}$$

By definition, $\overline{x}$ and x must be in opposite states for proper results.

One main purpose of the strobe is to disable the output of the gate during the interval that the inputs are changing; otherwise, false pulses, called *glitches,* can be produced. Other gates that use control inputs are the 7450, 7453, 7455, and 7462. Note that these strobes *do not* cause the outputs to have high-impedance states like three-state logic.

Conclusions

Strobed gates have control inputs that can disable the output while inputs are changing. Strobes are likely to be encountered when gate expanders are part of a circuit.

4

4000 Series CMOS Integrated Circuits

The TTL circuits described in Chap. 3 are not the only type of digital circuits in use. Another family of devices of growing importance is the complementary metal-oxide semiconductor (CMOS) circuit. While TTL gates are implemented using bipolar transistors, CMOS gates are based on field-effect transistors. Advantages of CMOS circuits include lower power and higher density (that is, more gates can be packed into one chip). On the other hand, TTL gates are faster in operation.

This chapter will emphasize the standard line of 4000 series CMOS circuits, which are manufactured by a number of companies. While not as extensive a line as the 5400/7400 TTL circuits, the 4000 series shares similar characteristics. These ICs can be readily interconnected with little concern for voltage levels, timing irregularities, or electrical interfacing. If proper procedures are followed, the CMOS and TTL gates can both be used in the same equipment. The CMOS gates are packaged as DIPs just like TTL.

HINTS FOR WORKING WITH CMOS CIRCUITS

There are two series of 4000 ICs available: the standard A type and the higher-voltage B type. The two are readily distinguished by the designator, such as in CD4000A and CD4001B. For these experiments, the B types of CMOS circuits are recommended. They offer a greater noise margin and hence provide more reliable operation. The type A CMOS can use a power supply voltage of 3 to 12 V, while type B accepts 3–18 V. In all cases, the experiments which follow will use a 5-V power supply. CMOS components are reliable and rugged, but special handling procedures should be followed.

Because of their high input impedance, CMOS devices are susceptible to damage from discharge of static electricity between any of the pins on the package. Such high voltage discharges may cause the gate oxide to break down. If you follow the suggestions given in this section, there should be little chance of failure from static electricity.

Power and ground leads should be connected to the dual in-line package (DIP) before any of the data inputs. Furthermore, the power supply should be turned on before any data inputs are applied. A reverse process should be followed for shutdown; first disable data inputs, and then turn off the power supply. (This sequence avoids the possibility of damaging the protective diodes in the chip.) Care should also be exercised in hooking up the power and ground wires so that the polarity of those signals are not reversed.

Short-circuiting the outputs of the gates should be prevented. Large capacitive loads on the outputs (in excess of 5 μF) act like a short circuit and must be avoided as well. Wired logic (wired AND or wired OR) configurations cannot be used with CMOS components. Paralleling inputs or outputs to gates is recommended only for gates in the same package.

Unused inputs should be connected to ground or to the power supply. The choice depends on the type of logic gate involved. Fan-out of CMOS gates can be very large. Typically, a fan-out of 50 is possible; however, each additional output connection slows the switching speed of the gate because of the increased capacitive load.

Table 4-1 4000 Series Gates

Type of Gate	Designator
Dual 3-input NOR with inverter	4000
Quad 2-input NOR	4001
Dual 4-input NOR	4002
Hex inverter	4009, 4049, 4069
Quad 2-input NAND	4011
Dual 4-input NAND	4012
Quad AND/OR	4019
Triple 3-input NAND	4023
Triple 3-input NOR	4025
Quad exclusive OR	4030, 4070
Quad 2-input OR	4071
Dual 4-input OR	4072
Triple 3-input AND	4073
Triple 3-input OR	4075
Quad exclusive NOR	4077
8-input OR/NOR	4078
Quad 2-input AND	4081
Dual 4-input AND	4082
AND-OR-invert	4085, 4086

The CMOS ICs are shipped on conductive foam carriers. They should remain on the foam whenever they are not inserted in the circuit. Aluminum foil can also be used on the work surface to ground the pins together if the foam is not available. Manufacturers recommend that table tops, tools, soldering irons, and even the experimenter be grounded. You can ground your hand with a metal megohm or conductive wrist strap that has a series 1-megohm ($M\Omega$) resistor to ground. Static electricity is more likely in dry weather, so extra care should be taken when humidity is low.

Examples of some of the gates available in the 4000 series are listed in Table 4-1. Many of the functions are the same as 5400/7400 ICs.

EXPERIMENT 4-1 IDENTIFYING THE OUTPUTS OF AN EQUALITY CIRCUIT

Purpose

To learn how to use 4000 series CMOS gates.

Parts

Item	Quantity
4071 quad OR	1
4081 quad AND	1
4049 hex inverter	1
SPDT switch	2
Logic probe	1

Procedure

Step 1.   The pin assignments for the three ICs are shown in Fig. 4-1. Wire the circuit as shown in Fig. 4-2a. Be sure to connect the ground and power supply pins first for each IC.

Step 2.   Ground both inputs. What is the output level? *(High)*

Step 3.   Switch input *A* high. Does the output change *(Yes; it becomes low)*

Step 4.   Ground input *A*. Switch input *B* high. How does the output respond? *(It remains low)*

Step 5.   Now switch both inputs low. What is the output? *(High)*

Ins and Outs

This circuit tests for the equality of the two inputs. In tabular form, the results you obtained would be as shown in Table 4-2.

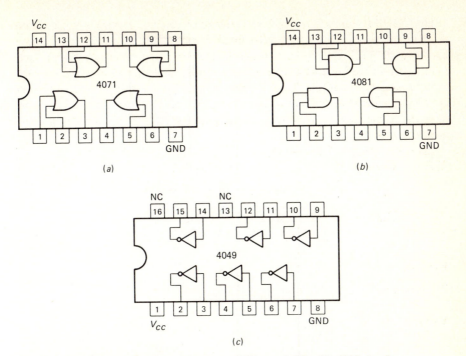

Fig. 4-1 Functional diagrams for the *(a)* 4071 quad OR, *(b)* 4081 quad AND, and *(c)* 4049 hex inverter.

Table 4-2   Equality Circuit Functions

Input		Output
A	*B*	
L	L	H
L	H	L
H	L	L
H	H	H

In the first and last rows, *A* and *B* are equal. In those cases the output level is high. When the inputs are in opposite status, the output is low.

The expression for the equality circuit is

$$AB + \overline{A}\,\overline{B}$$

This equation is sometimes written in the more compact form

$$A \odot B$$

Equality circuits are used in microprocessors and calculators. In Chap. 6 you will investigate the equality function further.

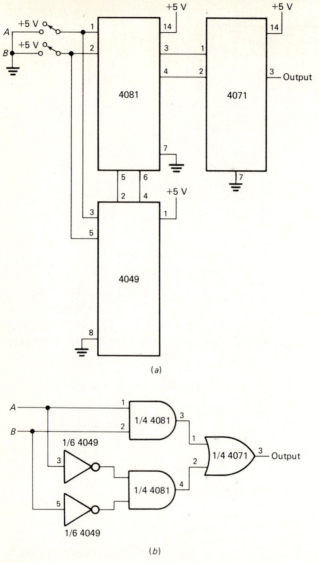

(a)

(b)

Fig. 4-2 The equality circuit produces a high-level output whenever the two inputs are equal. (*a*) Schematic; (*b*) logic diagram.

Conclusions

The 4000 series gates can be used to build logic networks just as the 5400/7400 TTL gates were used. To implement an equality circuit two inverters, two ANDs, and one OR function were used. This circuit generates a high output signal when the inputs are the same.

EXPERIMENT 4-2   MEASURING CMOS
PROPAGATION DELAY

Purpose

To compare the propagation delays in TTL and CMOS circuits.

Parts

Item	Quantity
4049 quad NAND	1
Oscilloscope	1

Procedure

Step 1.   Wire the circuit as shown in Fig. 4-3*a*. Refer to Fig. 4-1 for the pin assignment of the 4049 IC.

Step 2.   The procedure in this experiment repeats that of experiment measuring TTL propagation delays. You will want to refer back to the results you obtained there so that you can decide on the relative speeds of TTL and CMOS circuits.

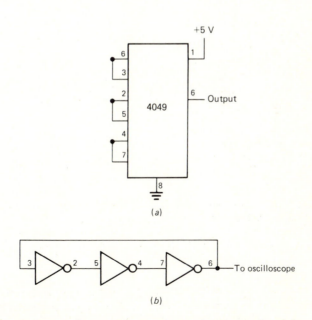

Fig. 4-3   This experiment will allow you to compare the propagation delay of CMOS with TTL inverters. *(a)* **Schematic;** *(b)* **logic diagram.**

Step 3. Using the oscilloscope, measure the delay in one CMOS inverter. Just as in Experiment 3-4, the delay is calculated by using the formula

$$\text{Delay} = \frac{\text{period of waveform}}{2 \times 3}$$

(About 100 ns)

Step 4. How much faster or slower is CMOS than TTL? *(TTL is about 10 times faster)*

Ins and Outs

The switching delay of the field-effect transistor is significantly longer than that of a bipolar transistor. This delay is reflected in the relative speeds of TTL and CMOS gates. Figure 4-4 shows another factor that affects the switching delay of CMOS gates. As the power supply voltage increases, the delay decreases. At a supply voltage of 5 V, the transition time is 100 ns for a load capacitance of 50 pF. That delay falls to 45 ns for if a 15-V power supply is used. This effect is similar to the speed-power product you saw characteristic of the different TTL subseries. As power consumption increases, speed also increases. In this case, tripling the power (because the voltage is multiplied by 3) causes a 55 percent gain in speed:

$$\frac{100 \text{ ns} - 45 \text{ ns}}{100} = 55 \text{ percent gain}$$

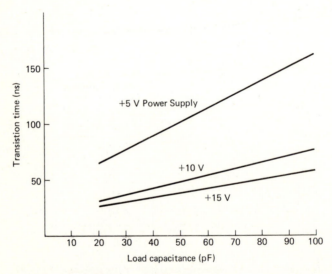

Fig. 4-4 As the supply voltage for CMOS gates increases, the transition time for switching between output levels decreases.

74

Conclusions

The CMOS gates can be run on low voltage and, therefore, low power. In such a situation, the switching time is relatively slow. By increasing the power, the transition time is decreased.

EXPERIMENT 4-3   INTERFACING CMOS TO TTL

Purpose

To use both CMOS and TTL gates in a common circuit.

Parts

Item	Quantity
4049 hex inverter	1
4081 quad AND	1
7432 quad OR	1
SPDT switch	1
Logic probe	1

Procedure

Step 1.   The circuit shown in Fig. 4-5 is identical to the equality circuit in Fig. 4-2a, except that a TTL gate is used for the final output. Connect the circuit as shown. Note that a double inversion is used on each output of the 4049. The 4049 is used to *buffer* the signals to the TTL gate.

Step 2.   Test this circuit to prove it performs the equality function by filling in Table 4-3.

Table 4-3   CMOS-TTL Equality Function

Input		Output
A	B	
L	L	(H)
L	H	(L)
H	L	(L)
H	H	(H)

Ins and Outs

The series *B* 4049 or 4050 buffer devices can sink the worst case current flowing out of up to two standard TTL gates. This means that the CMOS-TTL interface requires

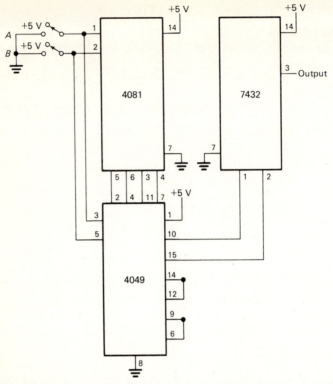

Fig. 4-5 The CMOS-TTL interface requires that the 4000 series device sink the TTL current. The series *B* devices used here can safely sink two TTL loads.

the use of one of these buffers on each CMOS output line. Table 4-4 lists the fan-out for the 4049 or 4050 buffers for other types of TTL gates.

Table 4-4 Minimum Fan-Out of CMOS Buffers to TTL Gates

TTL Subseries	4049/4050 Fan-Out
54/74	1
54H/74H	1
54L/74L	14
54S/74S	1
54LS/74LS	7

Conclusions

The interface of a CMOS gate to a TTL gate is facilitated with the 4049 and 4050 buffers. The 4049 inverts while the 4050 is noninverting. Each CMOS output line must be buffered before it is connected to the TTL gate input to guarantee an adequate capacity to sink the current from the TTL gate.

EXPERIMENT 4-4   INTERFACING TTL TO CMOS

Purpose

To demonstrate the method of connecting TTL gates to CMOS gates.

Parts

Item	Quantity
7409 quad AND	1
4071 quad OR	1
7404 hex inverter	1
1-kΩ resistor	2
SPDT switch	2
Logic probe	1

Procedure

Step 1.   Again, the equality circuit will serve to demonstrate the interfacing concepts. Connect the circuit shown in Fig. 4-6, omitting the resistors.

Step 2.   Try to verify the equality circuit voltage table. What do you find? *(Outputs always low)*

Step 3.   Insert the two 1-kΩ resistors.

Step 4.   Now test the equality circuit. Record your results in Table 4-5.

Table 4-5   TTL-CMOS Equality Circuit

Input		Output
A	*B*	
L	L	*(H)*
L	H	*(L)*
H	L	*(L)*
H	H	*(H)*

Ins and Outs

When interfacing TTL to CMOS with a common power supply, the minimum high-output voltage of 2.4 V from the TTL stage is lower than that needed for CMOS.

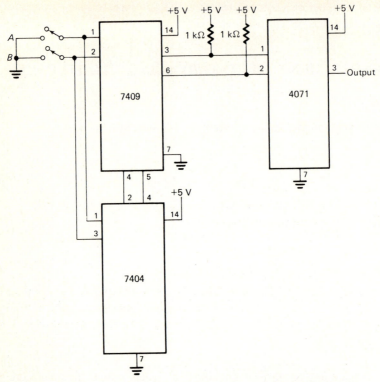

Fig. 4-6 **Pull-up resistors on open-circuit TTL gates are used to interface to CMOS gates.**

Figure 4-7 shows the relative voltage levels. An external resistor on the TTL outputs supplies the necessary active pullup to 3.5 V for CMOS. The minimum and maximum values for that resistor depend on the type of TTL gate used. These values are listed in Table 4-6. Note that open-collector TTL gates must be used for the interface.

Table 4-6   External Resistors for Interfacing TTL to CMOS

Type of TTL	Minimum R (Ω)	Maximum R (kΩ)
54/74	390	4.7
54H/74H	270	4.7
54L/74L	1500	27
54S/74S	270	4.7
54LS/74LS	820	12

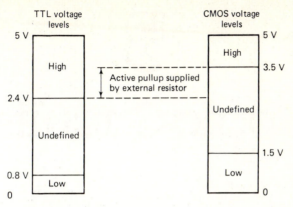

Fig. 4-7 Active pull-up resistors increase the minimum high-output level for TTL to that needed for CMOS.

Conclusions

By using open-collector TTL gates with pull-up resistors, the bipolar circuits can be connected to CMOS gates. The external resistor is necessary because the minimum voltage recognized as a high level by TTL is less than that required by CMOS.

A NOTE ON THE 74C SERIES CMOS

The low power consumption of CMOS can be combined with your circuit knowledge of the 7400 TTL series ICs. The 74C series of CMOS is a pin-for-pin and function-for-function replacement for the equivalent TTL component. The power dissipation per gate of the 74C ICs is on the order of 1.25 mW for low-frequency signals (<1 MHz) with a 5-V power supply. Propagation delay is about 50 ns. Manufacturers of the 74C series produce chips that can be interchanged with each other. About 40 of the most common TTL functions are available in their CMOS equivalents.

5

Logic Circuits

Now that you understand how the basic Boolean functions perform, you will be able to appreciate how they are combined into circuits that can do more complicated jobs. One of these tasks, which is repeatedly necessary, is being able to select one input out of many to forward to the remainder of the circuit. This selection processing can be accomplished by a *multiplexer*. Another interesting property of multiplexers is their ability to encode signals. As its name suggests, the *demultiplexer* is able to provide the service of delivering one input to any of several outputs.

Another series of ICs are the *multiple-bit comparators*. Based on the concepts that you are familiar with, the comparators are able to accept two numbers consisting of many binary digits and provide a signal to indicate which is larger. Finally, you will experiment with *triggers* that produce crisp, sharp pulses from slowly varying input waveforms. Such pulses are needed for reliable operation in digital equipment.

EXPERIMENT 5-1   CONSTRUCTING A MULTIPLEXER

Purpose

To connect a 1-to-4 multiplexing circuit.

Parts

Item	Quantity
7411 triple AND	2
7432 quad OR	1
SPDT switch	4
Logic probe	1

Procedure

Step 1.   Connect the circuit shown in the schematic of Fig. 5-1. After setting up the circuit, verify all connections by tying inputs A, B, C, and D (pin 1 on both 7411s and pins 3 and 10 of 7411 number 1) to 5 V. Put a logic probe on the output. For each of the conditions below, the output should remain high.

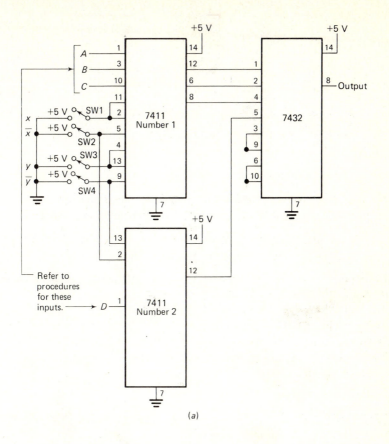

(a)

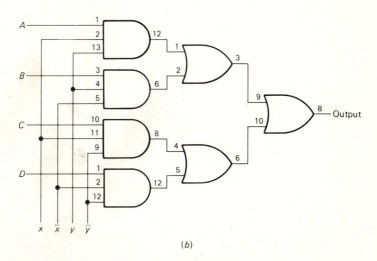

(b)

Fig. 5-1 The multiplexer selector inputs channel the desired data input to the output. (a) Schematic; (b) logic diagram.

Switch			
1	2	3	4
Closed	Open	Closed	Open
Closed	Open	Open	Closed
Open	Closed	Closed	Open
Open	Closed	Open	Closed

If your results do not agree, recheck all wiring.

Step 2.   Now ground input A, leaving the remaining inputs attached to the power supply. Close switches 1 and 3, open 2 and 4. What output level do you detect? *(Low)*

Step 3.   Next ground input B with all other inputs tied high. When switches 1 and 4 are closed and 2 and 3 open, what is the output? *(Low)*

Step 4.   Repeat the same procedure for inputs C and D using the switch patterns below.

	Grounded Input	Closed Switches	Open Switches
1.	C	2 and 3	1 and 4
2.	D	2 and 4	1 and 3

What are the outputs? *(1, low; 2, low)*

Step 5.   Summarize your results in Table 5-1.

Table 5-1   Switch Settings

Grounded Input	Switches 1 2 3 4	Output
A		*(Low)*
B		*(Low)*
C		*(Low)*
D		*(Low)*

Ins and Outs

The multiplexer is a device that has outputs which can be selected by placing the correct address on the switches. As the experiment demonstrates, the output results from the state of the switches as listed in Table 5-1. The action can be summarized as

$$A = 1 \text{ AND } 3$$
$$B = 2 \text{ AND } 3$$
$$C = 1 \text{ AND } 4$$
$$D = 2 \text{ AND } 4$$

(see Fig. 5-1*b*).

Conclusions

Most multiplexers can be characterized in the same way. To send any data input through the multiplexer, the correct address must be set on the selector inputs. The gate with ones on both selector lines will pass its input through. Multiplexers are often referred to by the short term MUX.

EXPERIMENT 5-2   WORKING WITH A DEMULTIPLEXER

Purpose

To demonstrate the operation of a digital demultiplexer.

Parts

Item	Quantity
7411 triple AND	2
SPDT switch	4
Logic probe	1

Procedure

Step 1.   Connect the circuit as shown in Fig. 5-2. With the input high, connect switches 1 and 3 to 5 V, leaving 2 and 4 grounded. Which output registers a high level? Did you find output 0 high? Figure 5-2*b* provides a logic diagram of this circuit. Verify your results.

Step 2.   If switches 2 and 3 were closed (5 V), 1 and 4 grounded, which output do you predict would change to a high level? Verify your prediction. *(Output 1)*

Step 3.   Continue the experiment with 1 and 4 closed, 2 and 3 grounded. What is the result? *(Output 2)*

Step 4.   Finally, with switches 1 and 3 grounded, close 2 and 4. Which output is addressed? *(Output 3)*

Step 5.   Complete Table 5-2 to record your results.

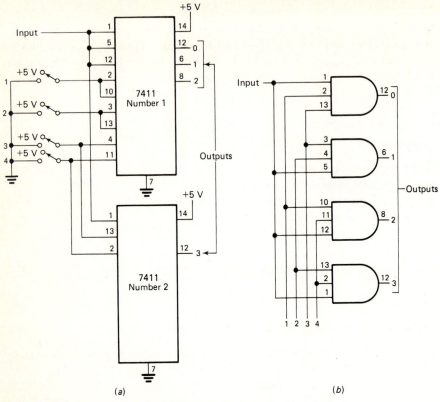

Fig. 5-2 The demultiplexer can deliver one input to any of several outputs.
(a) Schematic; (b) logic diagram.

Table 5-2   Demultiplexer

	Switch			
1	2	3	4	Output
Closed	Grounded	Closed	Grounded	(0)
Grounded	Closed	Closed	Grounded	(1)
Closed	Grounded	Grounded	Closed	(2)
Grounded	Closed	Grounded	Closed	(3)

Ins and Outs

Again, the AND gates, together with the switches, allow one signal to be routed to the correct destination. Demultiplexers select the AND gate with two ones on the address lines as the gate to produce an output.

Conclusions

In many ways a demultiplexer is like a multiplexer. Address or selector lines steer the signals along the indicated path. In general, the number of gates that can be controlled by n selector lines is $2^{n/2}$. In Experiments 5-1 and 5-2 the number of selector lines was 4, so the number of gates is $2^{4/2} = 2^2 = 4$.

EXPERIMENT 5-3   VERIFYING MULTIPLEXER FUNCTIONS

Purpose

To verify the truth table of an IC multiplexer.

Parts

Item	Quantity
74153 dual multiplexer	1
SPDT switch	3
Logic probe	1

Procedure

Step 1.   Using the schematic of Fig. 5-3 as a guide, wire the circuit. Notice that the two selector inputs actually form four lines into multiplexer (Fig. 5-3b).

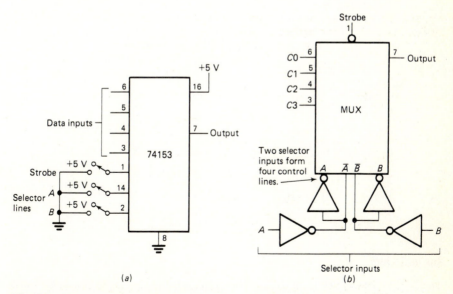

Fig. 5-3   The multiplexer splits the two selector inputs into four lines: A, $\bar{A}$, B, and $\bar{B}$. (a) Schematic; (b) logic diagram.

85

Step 2. The truth table for the 74153 4-to-1 multiplexer is given in Appendix A. First verify the action of the strobe input shown in Table 5-3. The Xs in the table are don't care conditions. Regardless of their value, the output remains the same.

Table 5-3 Strobe Control

Strobe	Selector Lines			Data Input				Output
	A	B		$C0$	$C1$	$C2$	$C3$	
H	X*	X		X	X	X	X	L

*X = don't care

Step 3. Now test the way in which input $C0$ is addressed, using Table 5-4 as a guide. After verifying that the output follows $C0$, you should try various combinations of the data inputs to test the don't care cases.

Table 5-4 $C0$ Input Addressing

Strobe	Selector Input			Data Input				Output
	A	B		$C0$	$C1$	$C2$	$C3$	
L	L	L		L	X	X	X	L
L	L	L		H	X	X	X	H

Step 4. Finally, check the remaining possibilities which route $C1$, $C2$, and $C3$ to the output (see Table 5-5).

Table 5-5 74153 Multiplexer Functions

Strobe	Selector Input			Data Input				Output
	A	B		$C0$	$C1$	$C2$	$C3$	
L	H	L		X	L	X	X	L
L	H	L		X	H	X	X	H
L	L	H		X	X	L	X	L
L	L	H		X	X	H	X	H
L	H	H		X	X	X	L	L
L	H	H		X	X	X	H	H

Ins and Outs

The 74153 contains two identical 1-to-4 multiplexers that operate in almost the same way as the one in Experiment 5-1. The only difference in the two multiplexers is the pin assignment for the inputs and outputs. The addresses for sending each data input to the output are

86

$$C0 = \overline{A}\,\overline{B}$$

$$C1 = A\,\overline{B}$$

$$C2 = \overline{A}\,B$$

$$C3 = A\,B$$

The multiplexers are enabled or disabled by the strobe input. The strobe must be low for the inputs to pass through.

Another useful function that the 74153 can perform is parallel-to-serial conversion. The 4-bit parallel data is applied to the data input lines, and then the data selector lines are sequenced to output one bit at a time. One pattern for the sequencing would be:

Selector Address		Output Bit	
$\overline{A}\,\overline{B}$	$C0$	Least-significant bit
$A\,\overline{B}$	$C1$	
$\overline{A}\,B$	$C2$	
AB	$C3$	Most-significant bit

Then repeat. Using this technique, the 4-bit quantity is delivered to the output a single bit at a time. The first bit sent is the least-significant bit (LSB). Can you think of a way to send the serial data with the most-significant bit (MSB) first? (*Reverse the selector address sequence.*)

Conclusions

The 74153 performs the same 1-to-4 multiplexer function as does the logic circuit in Experiment 5-1. The strobe must be low to enable the chip, and then the data selector lines can address each input line. The 1-to-4 multiplexer can also be used for parallel-to-serial conversion.

EXPERIMENT 5-4   USING THE 74155 DEMULTIPLEXER

Purpose

To review the principles of demultiplexer design.

Parts

Item	Quantity
74155 demultiplexer	1
SPDT switch	3
Logic probe	1

Procedure

Step 1. Connect the circuit in the schematic diagram of Fig. 5-4a.

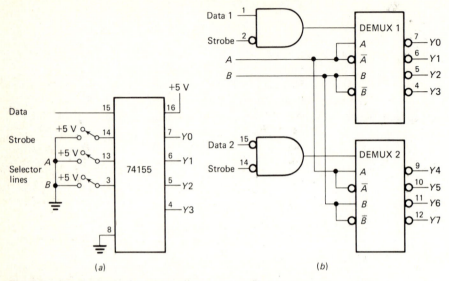

Fig. 5-4 The 74155 dual demultiplexers can be used separately or cascaded together. By various combinations of inputs, the device can be a 1-to-4 or 1-to-8 line demultiplexer. (a) Schematic; (b) internal logic diagram.

Step 2. Using the same techniques as in the previous experiment, verify the 1-to-4 line demultiplexer function table. Note that negative logic would be appropriate for this device.

Table 5-6   1-to-4 Demultiplexer Functions

Strobe	Selector Input A	B	Data Input	Y3	Y2	Y1	Y0
H	X	X	X	H	H	H	H
L	L	L	L	H	H	H	L
L	H	L	L	H	H	L	H
L	L	H	L	H	L	H	H
L	H	H	L	L	H	H	H
X	X	X	H	H	H	H	H

Step 3. By connecting the two demultiplexer data inputs and strobes together, a 1-to-8 demultiplexer can be formed. Wire pin 1 to pin 15 and pin 2 to pin 14. The output pins are then identified as

Pin Number	Output
7	$Y0$
6	$Y1$
5	$Y2$
4	$Y3$
9	$Y4$
10	$Y5$
11	$Y6$
12	$Y7$

Step 4. Test the operation of the demultiplexer. Does it operate as Table 5-7 indicates it should? In this case, the input used above for the strobe becomes the data input. The input used above for data becomes one of the selector address lines (see Fig. 5-4*b*).

Table 5-7   74155 1-to-8 Demultiplexer

Data Input* (Strobe)	Selector Input			Output							
	C†	B	A	$Y0$	$Y1$	$Y2$	$Y3$	$Y4$	$Y5$	$Y6$	$Y7$
H	X	X	X	H	H	H	H	H	H	H	H
L	L	L	L	L	H	H	H	H	H	H	H
L	L	L	H	H	L	H	H	H	H	H	H
L	L	H	L	H	H	L	H	H	H	H	H
L	L	H	H	H	H	H	L	H	H	H	H
L	H	L	L	H	H	H	H	L	H	H	H
L	H	L	H	H	H	H	H	H	L	H	H
L	H	H	L	H	H	H	H	H	H	L	H
L	H	H	H	H	H	H	H	H	H	H	L

*Pins 2 and 14 connected.
†Pins 1 and 15 connected.

Ins and Outs

The 1-to-4 demultiplexer uses the inputs associated with demultiplexer 2. Because both the input and outputs are inverted (indicated by the bubbles in Fig. 5-4*b*), the output and input levels are the same. Remember that an even number of inversions is equivalent to the original value.

To form a 1-to-8 demultiplexer, the two inputs and strobes are connected together. There is no inverter on the data line to DEMUX 1, so the two demultiplexers will always receive opposite values from their AND gates. When the AND for DEMUX 1 produces a one, for example, the AND for DEMUX 2 must supply a one. For this reason, using the data input for address selection is appropriate. Then the strobe actually determines when a low state will be read at the addressed output line.

Conclusions

The 74155 demultiplexer can be connected in configurations that supply 1-to-4 or 1-to-8 line demultiplexing. The cascading is accomplished by using a common strobe and data input.

EXPERIMENT 5-5   DECODING DECIMAL NUMBERS

Purpose

To verify the decoding of binary decimal digits.

Parts

Item	Quantity
7442 decoder	1
SPDT switch	4
Logic probe	1

Procedure

Step 1.   Connect the circuit as shown in Fig. 5-5.

Step 2.   With all inputs low, determine the states of the ten output lines. *(Pin 1 low, all others high)*

Step 3.   Next set the A input high, keeping B, C, and D low. Which output becomes high? *(Pin 2)*

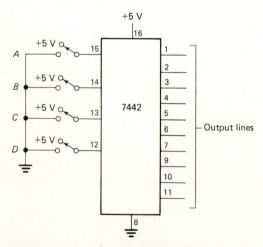

Fig. 5-5   The 7442 decoder changes the BCD input to an indication of the decimal digit by setting only the appropriate line high.

Step 4. Continue through the combination of inputs listed in Table 5-8.

Table 5-8   Decoder Inputs Function Table

Number	D	C	B	A	1	2	3	4	5	6	7	9	10	11
		Input						Output Pin Number						
0	L	L	L	L	L	H	H	H	H	H	H	H	H	H
1	L	L	L	H	H	L	H	H	H	H	H	H	H	H
2	L	L	H	L	H	H	L	H	H	H	H	H	H	H
3	L	L	H	H	H	H	H	L	H	H	H	H	H	H
4	L	H	L	L	H	H	H	H	L	H	H	H	H	H
5	L	H	L	H	H	H	H	H	H	L	H	H	H	H
6	L	H	H	L	H	H	H	H	H	H	L	H	H	H
7	L	H	H	H	H	H	H	H	H	H	H	L	H	H
8	H	L	L	L	H	H	H	H	H	H	H	H	L	H
9	H	L	L	H	H	H	H	H	H	H	H	H	H	L

Step 5. Convert the table to numeric values by assigning the values:

0 to a low input

1 to a high input

1 to a low output

0 to a high output

Your results should agree with those of Table 5-9.

Table 5-9   7442 Decoder Functions

Number	D	C	B	A	1	2	3	4	5	6	7	9	10	11
		Input						Output Pin Number						
0	0	0	0	0	1	0	0	0	0	0	0	0	0	0
1	0	0	0	1	0	1	0	0	0	0	0	0	0	0
2	0	0	1	0	0	0	1	0	0	0	0	0	0	0
3	0	0	1	1	0	0	0	1	0	0	0	0	0	0
4	0	1	0	0	0	0	0	0	1	0	0	0	0	0
5	0	1	0	1	0	0	0	0	0	1	0	0	0	0
6	0	1	1	0	0	0	0	0	0	0	1	0	0	0
7	0	1	1	1	0	0	0	0	0	0	0	1	0	0
8	1	0	0	0	0	0	0	0	0	0	0	0	1	0
9	1	0	0	1	0	0	0	0	0	0	0	0	0	1

Ins and Outs

The decimal digits from zero through nine can be represented by four binary digits as follows:

Decimal	Binary
0	0000
1	0001
2	0010
3	0011
4	0100
5	0101
6	0110
7	0111
8	1000
9	1001

This representation is called *binary-coded decimal* (BCD). The purpose of the 7442 is to change an input in BCD to its equivalent decimal number by lowering the output line that is assigned that value. All other output lines are left high.

By examining Table 5-9, you can see that the pin associated with each decimal digit is:

Pin	Digit
1	0
2	1
3	2
4	3
5	4
6	5
7	6
9	7
10	8
11	9

What happens if the input is larger than any decimal digit, for example, all inputs high, which is $1111_2 = 15_{10}$? Try this and other such undefined cases with the 7442 circuit. The result is that the outputs all remain high whenever the input is not within the proper range.

Conclusions

The 7442 is a decoder that converts the values on four input lines to 1 of 10 output lines. Hence it is called a 4-line-to-10-line decoder. The main purpose for this conversion is to translate BCD to decimal, so the device can also be referred to as a *BCD-to-decimal decoder*.

EXPERIMENT 5-6   COMPARING BCD NUMBERS

Purpose

To study the use of a magnitude comparator.

Parts

Item	Quantity
7485 magnitude comparator	1
SPDT switch	4
Logic probe	1

Procedure

Step 1.   Connect the circuit shown in Fig. 5-6a. The least-significant 2 bits of each input are tied high, so they are always equal.

Step 2.   Ground all the inputs except pin 15, which is switched to the high level. Which output is high? *(Pin 5)*.

Step 3.   Now ground pin 15 and place a high on pin 1 with the switch. How does the output change? *(Pin 7 becomes high)*

Step 4.   Notice that pin 5 is labeled "$A > B$" and pin 7 "$A < B$." When the input for bit $A3$ was high and $B3$ low, the output showed that the 4 bits of A were greater than B. A similar result occurred in step two with the size of the numbers reversed.

Step 5.   Repeat these experiments with pins 13 and 14. What do you find? *(A high on pin 13 produces a high output from pin 5; with the input of pin 14 high, pin 7 becomes high)*

Step 6.   Continue the experiment with various combinations of the inputs, noting each time which output is high. What happens when all inputs are the same? *(Pin 6 is high)*

Ins and Outs

As you have demonstrated, the 7485 compares and indicates the larger of the two inputs. Table 5-10 summarizes the functions of the 7485. For the time being, do not concern yourself with the cascading input columns of Table 5-10. As the table shows, the input quantities are compared in their MSBs first. If those bits are not equal, the

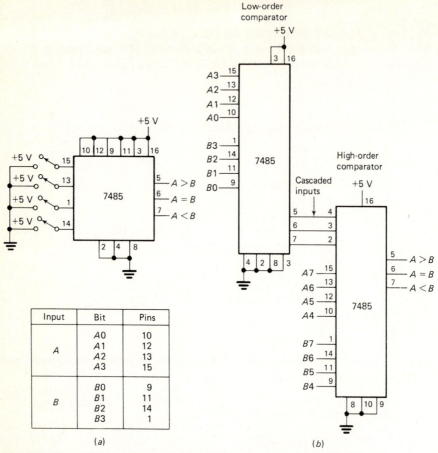

Input	Bit	Pins
A	A0	10
	A1	12
	A2	13
	A3	15
B	B0	9
	B1	11
	B2	14
	B3	1

(a)　　　　　　　　　　(b)

Fig. 5-6 Two 4-bit numbers can be compared with a magnitude comparator. Their relative size is indicated by the output signals. Larger numbers can be used as inputs by cascading the 7485. (a) Schematic; (b) cascaded comparators for 8-bit input.

proper output line becomes high. Otherwise, the next bits in order are matched. This process continues until the LSB is reached. If the two numbers are equal, the output depends only on the levels of the cascading inputs.

This set of cascading inputs is provided so that numbers longer than 4 bits can be compared. In fact, by cascading 7485 comparators, numbers with lengths of up to 120 bits can be used as inputs. A simpler case for only an 8-bit word length is shown in Fig. 5-6b. The cascading inputs are the key to the way these devices perform in this circuit.

The most-significant 4 bits of the A and B inputs are applied to the high-order 7485. If the two numbers differ in these bits, the outputs of the circuit (pins 5, 6, and 7 of the high-order comparator) will reflect the relative magnitudes. If bits 4 through 7 of the inputs are the same, the lower-order comparator determines the output. Bits 0 through 3 are examined, and the output pins of the low-order comparator indicate the result. If

Table 5-10   7485 Function Table

Comparison Input				Cascading Input			Output		
$A3, B3$	$A2, B2$	$A1, B1$	$A0, B0$	$A > B$	$A = B$	$A < B$	$A > B$	$A = B$	$A < B$
$A3 > B3$	X*	X	X	X	X	X	H	L	L
$A3 < B3$	X	X	X	X	X	X	L	L	H
$A3 = B3$	$A2 > B2$	X	X	X	X	X	H	L	L
$A3 = B3$	$A2 < B2$	X	X	X	X	X	L	L	H
$A3 = B3$	$A2 = B2$	$A1 > B1$	X	X	X	X	H	L	L
$A3 = B3$	$A2 = B2$	$A1 < B1$	X	X	X	X	L	L	H
$A3 = B3$	$A2 = B2$	$A1 = B1$	$A0 > B0$	X	X	X	H	L	L
$A3 = B3$	$A2 = B2$	$A1 = B1$	$A0 < B0$	X	X	X	L	L	H
$A3 = B3$	$A2 = B2$	$A1 = B1$	$A0 = B0$	H	L	L	H	L	L
$A3 = B3$	$A2 = B2$	$A1 = B1$	$A0 = B0$	L	H	L	L	H	L
$A3 = B3$	$A2 = B2$	$A1 = B1$	$A0 = B0$	L	L	H	L	L	H

*X = Don't care.

one number is larger, the output pins of the low-order comparator will show the result.

Note that the cascading inputs of $A > B$ and $A < B$ (pins 2 and 4, respectively) are grounded on that 7485. Furthermore, the cascading input for $A = B$ (pin 3) is high. From Table 5-10 you can see that if the low-order bits are equal, the output for $A = B$ (pin 6) will be the only high input.

Now examine how the output of the low-order comparator affects the high-order comparator. From Table 5-10 you see that these inputs are important only if bits 4 through 7 are equal. If that equality is the case, the cascading inputs to the high-order comparator decide the output levels. There are three possible situations (see Table 5-11).

Table 5-11

Output from Low-Order Comparator			Cascading Input Level for High-Order Comparator			Output from High-Order Comparator		
$A > B$	$A = B$	$A < B$	$A > B$	$A = B$	$A < B$	$A > B$	$A = B$	$A < B$
H	L	L	H	L	L	H	L	L
L	H	L	L	H	L	L	H	L
L	L	H	L	L	H	L	L	H

In the first case, bits $A0$ through $A3$ were greater than bits $B0$ through $B3$, so the output of the low-order comparator has a high $A > B$ output. This output becomes the input to the $A>B$ input. Because bits $A4$ through $B7$ were equal to $B4$ through $B7$, the high-order comparator output will agree with the cascading inputs. Therefore, the $A > B$ output is high. The other cascading inputs function in a similar manner if bits $B0$ through $B3$ are larger than $A0$ through $A3$, except that the $A < B$ outputs and cascading input are high. (*Row 3 in the table*)

What if the numbers are equal? Because the cascading input $A = B$ of the low-order comparator is tied to the supply voltage, that chip has a high on the $A = B$ output. That signal becomes the $A = B$ cascading input for the high-order 7485, thus forcing its $A = B$ output high also.

Conclusions

The 7485 magnitude comparator shows the relative size of two 4-bit inputs. By cascading them, these devices can compare even larger numbers. This latter capability is facilitated through the use of the cascading inputs. As long as the two numbers are not equal, these inputs are ignored. When the numbers are equal, the output is the result of the cascading input levels.

EXPERIMENT 5-7   SCHMITT TRIGGER

Purpose

To show the conversion of a slowly varying input to a square wave pulse train.

Parts

Item	Quantity
7414 Schmitt trigger inverter	1
1N456 diode	1
Signal generator	1
Oscilloscope	1

Procedure

Step 1.   Adjust the signal generator to produce a sine wave with a frequency of 60 to 100 Hz and a peak amplitude of 3 to 5 V. Observe the waveform on the oscilloscope by connecting the probe to point 1 on the circuit.

Step 2.   Connect the circuit as shown in Fig. 5-7a.

Step 3.   Observe the output waveform at point 2 on the scope. Draw a sketch of the waveform, showing the signal amplitude and period.

Step 4.   If you have a dual-trace slope, connect a probe to the signal converter for

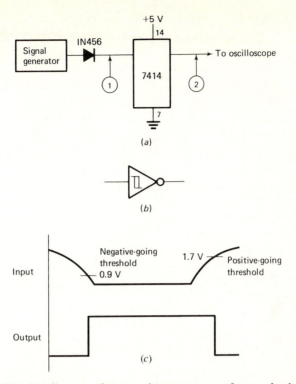

Fig. 5-7 **The Schmitt trigger produces a crisp square wave from a slowly varying input signal. Because the 7414 is an inverter, the output is 180° out of phase with the input.** *(a)* **Schematic;** *(b)* **symbol;** *(c)* **experimental waveforms.**

the upper trace and connect the Schmitt trigger output to the lower. What do you conclude about the phase relationship between the input and the output of the 7414? *(They are 180° out of phase)*

Ins and Outs

The Schmitt trigger (sometimes called a *regenerative comparator*) is the most commonly used device to produce sharp square-wave signals. The 7414 provides six Schmitt trigger inverters. It can convert not only sine waves, but also triangular, sawtooth, or arbitrary waveforms to crisp pulses.

The symbol for a Schmitt trigger inverter includes the rectangle with extended horizontal lines as shown inside the inverter in Fig. 5-7*b*. This shape is representative of the *hysteresis properties* of the trigger. The Schmitt trigger switches at one voltage level in going high and at a different voltage level on going low. The asymmetrical response is referred to as *hysteresis*, which is caused by a storage of energy in the device.

For the positive-going threshold voltage, the 7414 typically switches at 1.7 V. On the other hand, its negative-going threshold is about 0.9 V. The hysteresis (or difference) voltage is then 0.8 V. The hysteresis characteristics are shown in

Fig. 5-7c. The output becomes high when the input falls below the threshold value of 0.9 V. As the input becomes more positive, the switching of the output level does not occur until 1.7 V is reached.

Conclusions

One important application of Schmitt triggers is to change the slowly varying signal from a mechanical switch to a pulse input to digital circuits. Schmitt triggers are also combined with logic gates as in the 74113 dual NAND gates with Schmitt triggers.

The device is named for the inventor of the vacuum tube version of this circuit, which was originally used in television equipment. The only requirement placed on the input signal, which generates the output square wave, is that the input excursions exceed the limits of the hysteresis range. The input must have peak limits beyond both the positive and negative threshold voltages. The amplitude of the output is independent of the peak-to-peak voltage of the input.

EXPERIMENT 5-8   PRIORITY ENCODER

Purpose

To examine the functions of an encoder.

Parts

Item	Quantity
74147 priority encoder	1
LED	4
330-Ω resistor	4

Procedure

Step 1.   Connect the components as shown in Fig. 5-8. Note that in this experiment, if an LED glows, that output line is low. If an LED is off, that output is high.

Step 2.   What is the LED pattern with all inputs connected to the power supply? *(All off)*

Step 3.   Connect only pin 10 of the 74147 to ground. What is the pattern of the LEDs? *(A and D on, all others off)*

Step 4.   Continue to connect the inputs to ground in the order shown in Table 5-12. Write your results in the space provided. Compare them to the expected values.

Step 5.   Now connect pins 10 and 5 to ground. Do the same with pins 5 and 4. Continue to try combinations of two, three, or four input pins. What do you conclude? *(The pin nearest the top of Table 5-12 controls the output when multiple input pins are tied to ground)*

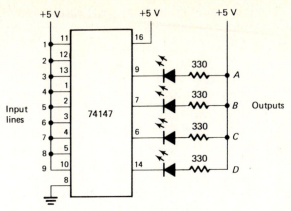

Note: Output *A* is the LSB, output *D* is the MSB.

Fig. 5-8 The 74147 priority encoder selects the highest number input line and provides its BCD digit on the output pins.

Table 5-12   Encoder Results

High-Input Pin	Your Results				Expected Results			
	D	*C*	*B*	*A*	*D*	*C*	*B*	*A*
None					H	H	H	H
10					L	H	H	L
5					L	H	H	H
4					H	L	L	L
3					H	L	L	H
2					H	L	H	L
1					H	L	H	H
13					H	H	L	L
12					H	H	L	H
11					H	H	H	L

Ins and Outs

The 74147 not only converts the selected line to its BCD equivalent, but also gives highest priority to the largest digit. Table 5-13 presents a complete functional description of this encoder. When no input is selected (all high), the outputs are all high, equivalent to a 0000 BCD output if we let the high level be represented by a zero. (The low level then stands for a one.)

As the functional table (Table 5-13) shows, when there is no input grounded, the BCD output is zero. When input digit 9 is grounded, the BCD output is 1001. If the pin for digit 8 is grounded, the output is 1000, and so on. Another aspect of the 74147

Table 5-13 74147 Function Table

Input									Output							
Pin 10	5	4	3	2	1	13	12	11	Voltage Level				BCD			
Digit 9	8	7	6	5	4	3	2	1	D	C	B	A	D	C	B	A
H	H	H	H	H	H	H	H	H	H	H	H	H	0	0	0	0
L	X	X	X	X	X	X	X	X	L	H	H	L	1	0	0	1
H	L	X	X	X	X	X	X	X	L	H	H	H	1	0	0	0
H	H	L	X	X	X	X	X	X	H	L	L	L	0	1	1	1
H	H	H	L	X	X	X	X	X	H	L	L	H	0	1	1	0
H	H	H	H	L	X	X	X	X	H	L	H	L	0	1	0	1
H	H	H	H	H	L	X	X	X	H	L	H	H	0	1	0	0
H	H	H	H	H	H	L	X	X	H	H	L	L	0	0	1	1
H	H	H	H	H	H	H	L	X	H	H	L	H	0	0	1	0
H	H	H	H	H	H	H	H	L	H	H	H	L	0	0	0	1

operation shown by Table 5-13 is that if two or more inputs are grounded, the output only reflects the highest digit.

Given this type of operation, how might the 74147 be used? One popular application is keyboard encoding. If the numeric keyboard generates a ground when the button is depressed, that signal routed to the 74147 will produce the BCD value of the numeric key that was selected. Only the highest-order digit will be encoded if more than one key is pressed.

Conclusions

The 74147 can be used to convert the digital keys on a keyboard to their BCD equivalent. Because the inputs are ranked in priority order, only the highest digit is accepted if multiple keys are pressed.

6

Arithmetic and Logic Circuits

The next group of functions are probably the first to come to mind in any consideration of digital circuitry. After all, the reason for using a digital form of information is so that it can be manipulated. These circuits make it possible to add, subtract, multiply, and divide digital quantities. But they may also be used to transform the values through Boolean operations.

You have used the binary number system in some of the earlier experiments. In this chapter, you will see how arithmetic operations are performed on binary numbers. Other number systems are also in use. You are quite familiar with the decimal numbers and also have used BCD numbers. Other numbering systems often used with digital circuits are the base 8, or *octal,* and base 16, or *hexadecimal,* systems. These latter two systems are simply a shorter way of writing the binary numbers. This effect is apparent in Table 6-1, which lists the values of numbers in the different systems. Also noteworthy is the use of letters to represent numbers in the hexadecimal system. You will want to refer to this table frequently to convert from one number system to another.

To distinguish one number system from another, a subscript is appended to indicate the base. For example,

101_2 Binary

20_8 Octal

14_{10} Decimal

14_{16} Hexadecimal

Note that in the last two examples, confusion between the numbers would result if the subscripts were omitted. To convert one number system to another, simply locate it in the proper column of the table, and then read the equivalent for another base in the same row. For example,

$$36_{10} = 10\ 0100_2 = 44_8 = 24_{16}$$

Table 6-1 Number System Equivalents

Decimal	Binary	Octal	Hexadecimal	Decimal	Binary	Octal	Hexadecimal
0	0	0	0	33	10 0001	41	21
1	1	1	1	34	10 0010	42	22
2	10	2	2	35	10 0011	43	23
3	11	3	3	36	10 0100	44	24
4	100	4	4	37	10 0101	45	25
5	101	5	5	38	10 0110	46	26
6	110	6	6	39	10 0111	47	27
7	111	7	7	40	10 1000	50	28
8	1000	10	8	41	10 1001	51	29
9	1001	11	9	42	10 1010	52	2A
10	1010	12	A	43	10 1011	53	2B
11	1011	13	B	44	10 1100	54	2C
12	1100	14	C	45	10 1101	55	2D
13	1101	15	D	46	10 1110	56	2E
14	1110	16	E	47	10 1111	57	2F
15	1111	17	F	48	11 0000	60	30
16	1 0000	20	10	49	11 0001	61	31
17	1 0001	21	11	50	11 0010	62	32
18	1 0010	22	12	51	11 0011	63	33
19	1 0011	23	13	52	11 0100	64	34
20	1 0100	24	14	53	11 0101	65	35
21	1 0101	25	15	54	11 0110	66	36
22	1 0110	26	16	55	11 0111	67	37
23	1 0111	27	17	56	11 1000	70	38
24	1 1000	30	18	57	11 1001	71	39
25	1 1001	31	19	58	11 1010	72	3A
26	1 1010	32	1A	59	11 1011	73	3B
27	1 1011	33	1B	60	11 1100	74	3C
28	1 1100	34	1C	61	11 1101	75	3D
29	1 1101	35	1D	62	11 1110	76	3E
30	1 1110	36	1E	63	11 1111	77	3F
31	1 1111	37	1F	64	100 0000	100	40
32	10 0000	40	20				

EXPERIMENT 6-1 TESTING HALF-ADDERS

Purpose

To obtain practice in binary addition.

Parts

Item	Quantity
7408 quad AND	1
7486 quad exclusive OR	1
7405 hex inverter	1
330-Ω resistor	2
LEDs	4
SPDT switch	2

Procedure

Step 1. Wire the circuit shown in Fig. 6-1a. Be sure to note that the LEDs attached to the power supply have polarity reverse from those which are grounded.

Step 2. Test your circuit by connecting both inputs to 5 V. All LEDs except the one attached to pin 10 of the 7405 should glow.

Step 3. Let a 5-V input represent a one and a 0-V input a zero. Fill in Table 6-2 with the results that you obtain from the various input combinations for this circuit. *(If a LED is on, the pin is high)*

Table 6-2

Input		Output	
A	B	Sum	Carry
0	0	*(0)*	*(0)*
0	1	*(1)*	*(0)*
1	0	*(1)*	*(0)*
1	1	*(0)*	*(1)*

Step 4. The binary addition table is given below. What do you conclude about your circuit?

0	0	1	1
+0	+1	+0	+1
0	1	1	10

$^{\llcorner}$Carry

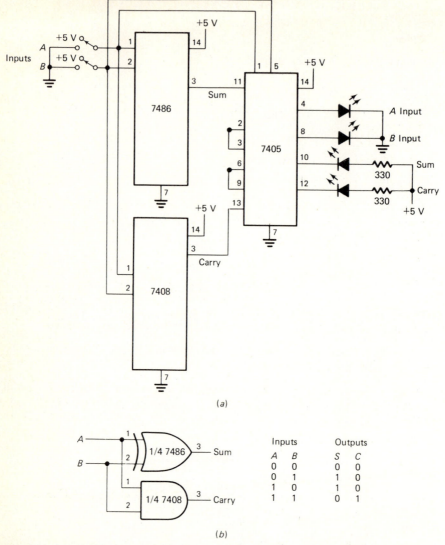

(a)

(b)

Fig. 6-1 **A half-adder produces the sum and carry bits from the two binary inputs.** (a) **Schematic;** (b) **half-adder logic diagram.**

(*The circuit performs binary addition correctly, except when a carry is generated. Then 2 bits are needed to store the sum*) Do not dismantle this circuit, as you will use part of it again in the next experiment.

Ins and Outs

Normally, addition is required on numbers consisting of more than one digit. Because no *carry in* from the digit to the right is provided, this circuit is called a *half-adder*. In decimal arithmetic, addition also produces carries

$$\text{Carry} \quad 1$$

$$19_{10}$$

$$+ \ 7_{10}$$

$$\overline{26_{10}}$$

When 9 and 7 are added, the sum is 6 "with one to carry." Exactly the same process is used for binary arithmetic.

The sum in this adder can be written as

$$S = A \oplus B$$

with the exclusive OR producing the correct digit. The carry is needed only when both inputs are ones. In other words, the carry is one if A AND B are one:

$$C = AB$$

As the logic diagram in Fig. 6-1b shows, the gates just implement these equations. The inverters and the LEDs are a handy way to show when the inputs and the outputs are high or low. The carry line is often called the *carry out*, or just C_o, because it is an output of the circuit.

Conclusions

The half-adder calculates the sum of the 2 input bits and sets the carry line, if necessary. The half-adder does not provide for an input carry from the next-lower digit.

EXPERIMENT 6-2   WORKING WITH A FULL-ADDER

Purpose

To demonstrate complete binary addition.

Parts

Item	Quantity
7408 quad AND	1
7432 quad OR	1
7486 quad exclusive OR	1
SPDT switch	3
Logic probe	1

Procedure

Step 1. Connect the circuit as shown in Fig. 6-2a. The inputs to the adder include A and B as well as the carry in C_i.

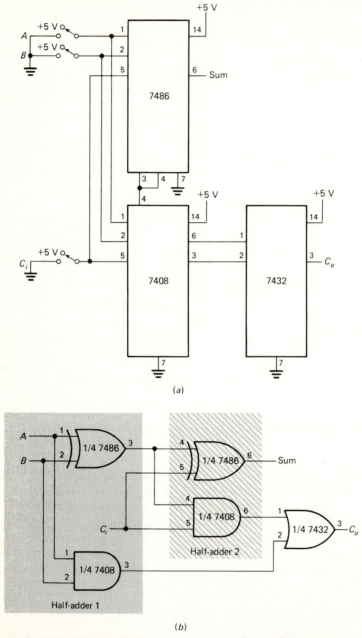

(a)

(b)

Fig. 6-2 The full-adder accepts the carry in from the next-lower bit position and produces a carry out to the next-higher bit adder. *(a)* Schematic; *(b)* logic diagram.

Step 2. Ground the carry in. Repeat adding the various combinations of A and B just like you did in the previous experiment. Record your results in Table 6-3.

Table 6-3   Full-Adder with $C_i = 0$

Input		Output	
A	B	Sum	C_o
0	0	(0)	(0)
0	1	(1)	(0)
1	0	(1)	(0)
1	1	(0)	(1)

Step 3. Now perform the same series of additions as in Step 2, but connect C_i to the 5-V supply first. Write down your answers in Table 6-4.

Table 6-4   Full-Adder with $C_i = 1$

Input		Output	
A	B	Sum	C_o
0	0	(1)	(0)
0	1	(0)	(1)
1	0	(0)	(1)
1	1	(1)	(1)

Step 4. Compare the outcome of Step 2 with that of the half-adder. How do they compare? (*You should have found them to be the same*) Do not disassemble this circuit. It is used in the next experiment.

Ins and Outs

The full-adder is needed to add numbers of more than 1 bit in length. You will have an opportunity to examine the capability in detail in following experiments. For now let us review the operation of the test circuit. The binary addition table can be extended to take into account the value of C_i as shown in Table 6-5. The full-adder varies from the half-adder only in the bottom part of the table when the carry in is nonzero.

The full-adder is constructed from two half-adders as shown in Fig. 6-2b. Half-adder 1 performs the operation of summing inputs A and B. The second half-adder combines the carry in with the partial sum to produce the final sum and the carry out.

Conclusions

The full-adder differs in operation from the half-adder only when the carry in bit is one. In such cases, this adder generates a corrected sum and a carry out. The output carry bit becomes the input carry to the next higher stage.

Table 6-5   Full-Adder Summary

	Input		Output	
	A	B	S	C_o
$C_i = 0$	0	0	0	0
	0	1	1	0
	1	0	1	0
	1	1	0	1
$C_i = 1$	0	0	1	0
	0	1	0	1
	1	0	0	1
	1	1	1	1

EXPERIMENT 6-3   DETERMINING THE OUTPUTS OF A 2-BIT ADDER

Purpose

To connect full-adders in a multiple-bit configuration.

Parts

Item	Quantity
7408 quad AND	1
7432 quad OR	1
7486 quad exclusive OR	1
SPDT switch	4
Logic probe	1

Procedure

Step 1.   Using the circuit in Experiment 6-2 as the basis for the bit 0 adder, wire the two full-adders as shown in Fig. 6-3. (Remove the switch from the carry in circuit and ground the C_i input for bit 0. That switch then becomes available for one of the inputs to the bit 1 adder. Wire the second adder, using the unused gates of the ICs.)

Step 2.   Test your adder with zero inputs for both bits of A and B. All outputs should be zero. Repeat with all inputs equal to one. You should find

$$S0 = 0$$
$$S1 = 1$$
$$C_o 0 = 1$$
$$C_o 1 = 1$$

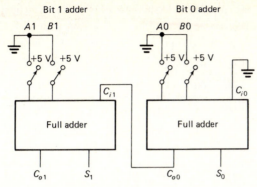

Fig. 6-3 A 2-bit adder is formed from two full-adders. The carry out of the lower bit becomes the carry in for the upper bit.

Step 3. After confirming the operation, test your adder with the inputs listed in Table 6-6.

Ins and Outs

The 2-bit adder is constructed by cascading full-adders. The carry out from the lower bit becomes the carry in to the higher bit. Because each carry in may change the

Table 6-6   2-Bit Adder Function Table

Input				Measured Output				Expected Output			
$B1$	$B0$	$A1$	$A0$	C_o1	C_o0	$S1$	$S0$	C_o1	C_o0	$S1$	$S0$
0	0	0	0					0	0	0	0
0	0	0	1					0	0	0	1
0	0	1	0					0	0	1	0
0	0	1	1					0	0	1	1
0	1	0	0					0	0	0	1
0	1	0	1					0	1	1	0
0	1	1	0					0	0	1	1
0	1	1	1					1	1	0	0
1	0	0	0					0	0	1	0
1	0	0	1					0	0	1	1
1	0	1	0					1	0	0	0
1	0	1	1					1	1	0	1
1	1	0	0					0	0	1	1
1	1	0	1					1	1	0	0
1	1	1	0					1	0	0	1
1	1	1	1					1	1	1	0

next-higher sum, this circuit is called a *ripple carry* adder. The operation of the adder as listed in Table 6-6 can be verified by comparison with all 2-bit sums.

A	00	01	10	11		00	01	10	11
B	00	00	00	00		01	01	01	01
SUM	00	01	10	11		01	10	11	100
A	00	01	10	11		00	01	10	11
B	10	10	10	10		11	11	11	11
SUM	10	11	100	101		11	100	101	110

For 3-bit sums, the $C_o 1$ bit serves as the MSB.

Conclusions

Multiple-bit adders use cascaded full-adders. The length of the sum is essentially unlimited. The time for the carries to propagate through the sum in a ripple carry adder increases with the number of digits.

EXPERIMENT 6-4   USING INTEGRATED-CIRCUIT ADDERS

Purpose

To become familiar with the 7482 adder.

Parts

Item	Quantity
7482 adder	1
LEDs	3
SPDT switch	4
Logic probe	1

Procedure

Step 1.   Connect the circuit shown in Fig. 6-4. Ground the input carry, pin 5. This low voltage is equivalent to a zero input. NOTE: The ground and power pins of the 7482 do not follow the normal numbering conventions. Pin 11 is ground and pin 4 power.

Step 2.   Using the switches, complete the function table given in Table 6-7.

110

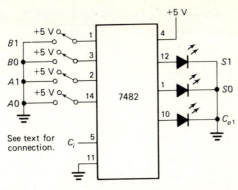

Fig. 6-4 The 7482 adder is an implementation of the circuit built in Experiment 6-5. This adder provides the sum of each bit and the carry from the high-order bits as outputs.

Table 6-7 7482 Function Table with $C_i 0 = 0$

Input				Measured Output			Expected Output		
A0	B0	A1	B1	S0	S1	$C_o 1$	S0	S1	$C_o 1$
0	0	0	0				0	0	0
1	0	0	0				1	0	0
0	1	0	0				1	0	0
1	1	0	0				0	1	0
0	0	1	0				0	1	0
1	0	1	0				1	1	0
0	1	1	0				1	1	0
1	1	1	0				0	0	1
0	0	0	1				0	1	0
1	0	0	1				1	1	0
0	1	0	1				1	1	0
1	1	0	1				0	0	1
0	0	1	1				0	0	1
1	0	1	1				1	0	1
0	1	1	1				1	0	1
1	1	1	1				0	1	1

Step 3. Now connect the input carry, pin 5, to 5 V.

Step 4. Using the switches, create the input conditions listed in Table 6-8. Record your results. Save the circuit. It will be used in the next experiment.

Table 6-8   7482 Function Table with $C_i 0 = 1$

Input				Measured Output			Expected Output		
$A0$	$B0$	$A1$	$B1$	$S0$	$S1$	C_o1	$S0$	$S1$	C_o1
0	0	0	0				1	0	0
1	0	0	0				0	1	0
0	1	0	0				0	1	0
1	1	0	0				1	1	0
0	0	1	0				1	1	0
1	0	1	0				0	0	1
0	1	1	0				0	0	1
1	1	1	0				1	0	1
0	0	0	1				1	1	0
1	0	0	1				0	0	1
0	1	0	1				0	0	1
1	1	0	1				1	0	1
0	0	1	1				1	0	1
1	0	1	1				0	1	1
0	1	1	1				0	1	1
1	1	1	1				1	1	1

Ins and Outs

The function tables for the 7482 are the same as those for the 2-bit adder you constructed for the previous experiment. Because the input carry was grounded in that circuit, it corresponds to a 7482 with $C_i = 0$. By rearranging Table 6-6 slightly, you can see its equivalence with Table 6-7.

The purpose of the input carry on the 7482 is to allow cascading many of the devices for construction of wider adders. The 7482 also is characterized by a ripple, or a serial carry. Such adders can be used for medium-speed applications. The low-order carry is connected internally to the adder. Other adders are available in 1-bit and 4-bit configurations.

Conclusions

Building adders from logic gates is unnecessary because they are produced in MSI configurations with costs as low as three times that of an SSI chip. The time saved in not having to make all the connections of simpler gates and reduced space on the circuit board more than compensates for the higher price when adders are called for in digital equipment.

EXPERIMENT 6-5   CONVERTING ADDERS
TO SUBTRACTORS

Purpose

To verify that adders can also be used to subtract.

Parts

Item	Quantity
7405 inverter	1
7482 adder	1
LEDs	3
SPDT switch	4
Logic probe	1

Procedure

Step 1.   The same circuit as used in Experiment 6-4 is used again, but inverters are inserted in the B input lines. Make the connections as shown in Fig. 6-5.

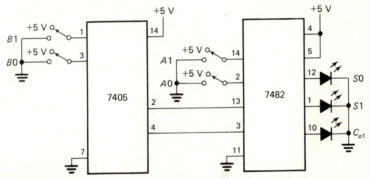

Fig. 6-5   By inverting one of the input numbers and providing a carry in equal to one, the adder becomes a subtractor.

Step 2.   Connect the carry in of the adder (pin 5) to 5 V.

Step 3.   Verify that subtraction is correctly performed by the circuit by letting $B = 1$ and $A = 1$, that is, $A0 = 1$, $A1 = 0$, $B0 = 1$, and $B1 = 0$. Did your answer equal zero? (*Ignore the output carry bit*)

Step 4.   Try again with the numbers given below. Is the sum zero in each case?

A	$A0$	$A1$	B	$B0$	$B1$
2	0	1	2	0	1
3	1	1	3	1	1

113

Step 5. What happens if you let $A0 = 1$, $A1 = 1$, $B0 = 1$, and $B1 = 0$? Is the answer $S0 = 0$ and $S1 = 0$? Is that correct?

Ins and Outs

The subtracter that you are using is based on a coded form of binary numbers called the *2's complement* system. The reason for inventing complement systems is to represent both positive and negative quantities in binary. Table 6-9 shows how the numbers are interpreted in a 2's complement system of 2-bit length.

Table 6-9   2's Complement

	Binary	Decimal	Meaning
Positive numbers	01	1	1
	00	0	0
Negative numbers	11	3	−1
	10	2	−2

In Step 5 the numbers involved were

$$A = 11_2 = 3_{10}$$
$$B = 01_2 = 1_{10}$$

which translate to

$$A = -1$$
$$B = 1$$

in 2's complement notation. If B is subtracted from A, the answer you expect is -2. Using the complement table in the opposite direction, the 2's complement of -2 is 10_2—exactly the value of the sum bits produced in Step 5.

Two-bit numbers do not produce very interesting results for subtraction, but if the same idea is extended to longer bit lengths, very useful operations can be created. In fact, every digital computer uses the same process with typical bit lengths of 8, 16, or 32. Consider the 3-bit 2's complement number system shown in Table 6-10. The range of values that we can work with has been doubled.

Distinguishing positive from negative 2's complement numbers is quite easy. Just look at the MSB. That digit is the *sign bit*. The sign bit for positive numbers is always zero and for negative numbers, always one in a 2's complement system. For completeness, you should know that there is another system, *1's complement*, that is also used to symbolize negative and positive numbers. This system is less used than 2's complement.

Table 6-10 A 3-Bit 2's Complement Number System

	2's Complement	Decimal Equivalent
Positive numbers	011	3
	010	2
	001	1
	000	0
Negative numbers	111	−1
	110	−2
	101	−3
	100	−4

Conclusions

Adders become subtracters by inverting all bits of one input and making the lowest-order carry in equal to one. Complement number systems are needed to provide positive and negative number identification. The sign bit in such number systems is zero for positive and one for negative quantities.

EXPERIMENT 6-6 VERIFYING LOGIC FUNCTIONS OF AN ALU

Purpose

To gain experience with ALUs.

Parts

Item	Quantity
74181 ALU	1
SPDT switch	8
LEDs	4
Logic probe	1

Procedure

Step 1. Because the 74181 ALU can do many jobs based on its *programming,* it takes a little study of its organization to gain confidence in its use. Table 6-11 is a grouping of its pins by functions. The group 1 functions are just the 4-bit quantities that are the data inputs A and B and a carry in bit. The group 2 pins program the ALU to carry out addition, subtraction, or Boolean functions. The mode control specifies logic (Boolean) operations if high and arithmetic operations if low. Because this experiment concerns logic operations, M (pin 8) is high in Fig. 6-6.

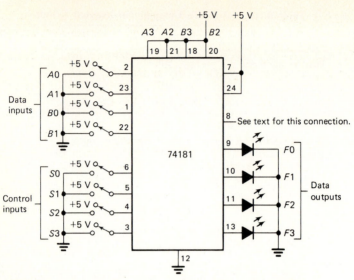

Fig. 6-6 The 74181 is a 4-bit ALU that can execute either Boolean or arithmetic functions, depending on the programming of its control inputs.

Table 6-11 Signal Groups for the 74181

Group	Purpose	Designation	Pins
1	Data input A	$A3, A2, A1, A0$	19, 21, 23, 2
	Data input B	$B3, B2, B1, B0$	18, 20, 22, 1
	Ripple carry in	C_n	7
2	Function selection S	$S3, S2, S1, S0$	3, 4, 5, 6
	Mode control (arithmetic or Boolean)	M	8
3	Data outputs F	$F3, F2, F1, F0$	13, 11, 10, 9
	Comparator output	$A = B$	14
	Carry output	P	15
	Ripple carry output	C_{n+4}	16
	Carry generate output	G	17

The third group of pins are outputs. The F pins provide the answers. A comparator is built into the ALU, so whenever the A input is equal to the B input, the $A = B$ (pin 14) output is high. (This operation is independent of the F outputs.) Several forms of the output carry are available for cascading ALUs and for inputs to carry generators (such as the 74182).

Step 2. Connect the circuit as shown in Fig. 6-6. Note that the upper 2 bits of both inputs are tied high, thus making them all equal. The carry in is low. NOTE: Use extra care to avoid bending pins on the 24-pin DIP when inserting or removing it.

Step 3. The logic functions that the ALU provides are listed in Table 6-12. The steps that follow will assist you in verifying the table entries.

Table 6-12 74181 ALU Logic Functions*

Function Selection				
$S3$	$S2$	$S1$	$S0$	Operation (Value of F)
0	0	0	0	$\overline{A}$
0	0	0	1	$\overline{A + B}$
0	0	1	0	$\overline{A} B$
0	0	1	1	0
0	1	0	0	$\overline{AB}$
0	1	0	1	$\overline{B}$
0	1	1	0	$A \oplus B$
0	1	1	1	$A \overline{B}$
1	0	0	0	$\overline{A} + B$
1	0	0	1	$\overline{A \oplus B}$
1	0	1	0	B
1	0	1	1	AB
1	1	0	0	1
1	1	0	1	$A + \overline{B}$
1	1	1	0	$A + B$
1	1	1	1	A

*When input M is high.

Step 4. To generate NOT A, inputs are as follows:

$$A0 = 1 \qquad S0 = 0$$
$$A1 = 0 \qquad S1 = 0$$
$$B0 = X \qquad S2 = 0$$
$$B1 = X \qquad S3 = 0$$

Where X means don't care (remember that bits $A2$ and $A3$ are ones). The F outputs you should obtain are

$$F0 = 0 \qquad\qquad F2 = 0$$
$$F1 = 1$$
$$\qquad\qquad F3 = 0$$

These outputs are the 1's complement of A. That is, each bit of A is inverted to form its 1's complement. This operation may be written

$$F = \overline{A}$$

F equals NOT A, in other words.

Step 5. NOR.

	Input		Expected Output
$A0 = 0$		$S0 = 1$	$F0 = 0$
$A1 = 0$		$S1 = 0$	$F1 = 0$
$B0 = 1$		$S2 = 0$	$F2 = 0$
$B1 = 1$		$S3 = 0$	$F3 = 0$

Recall that the upper bits of both input quantities are tied high.

The inputs and expected outputs for the remaining steps are listed in Table 6-13. As you perform each step, compare your answer to the one given. Do not take the circuit apart. It is used in the following experiment.

Table 6-13   Verifying Logic Operations

Step	Input								Expected Output				Function
	$A1$	$A0$	$B1$	$B0$	$S3$	$S2$	$S1$	$S0$	$F3$	$F2$	$F1$	$F0$	
6	1	1	0	0	0	0	1	0	0	0	0	0	A NOT AND B
7	X	X	X	X	0	0	1	1	0	0	0	0	Zero
8	0	1	1	1	0	1	0	0	0	0	0	1	NAND
9	X	X	0	0	0	1	0	1	0	0	1	1	NOT B
10	1	0	0	1	0	1	1	0	0	0	1	1	Exclusive OR
11	1	0	0	0	0	1	1	1	0	0	1	0	A AND B NOT
12	0	0	1	1	1	0	0	0	1	1	1	1	A NOT OR B
13	0	0	0	0	1	0	0	1	1	1	1	1	Exclusive NOR
14	X	X	0	0	1	0	1	0	1	1	0	0	B
15	1	0	1	1	1	0	1	1	1	1	1	0	AND
16	X	X	X	X	1	1	0	0	0	0	0	1	One
17	0	0	0	1	1	1	0	1	1	1	1	0	A OR B NOT
18	0	0	1	1	1	1	1	0	1	1	1	1	OR
19	1	1	X	X	1	1	1	1	1	1	1	1	A

Ins and Outs

As you have demonstrated, the ALU can execute AND, OR, NAND, NOR, NOT, and more complicated functions just by changing the values of the function selection bits. The correctness of your answers can best be proven with direct application of Boolean algebra.

Step	Proof
4	$A = 1101, \bar{A} = 0010$
5	$A = 1100, B = 1111, \bar{A} + \bar{B} = 0000$
6	$A = 1111, B = 1100, \bar{A}B = 0000$
7	Zero output
8	$A = 1110, B = 1111, \overline{AB} = 0001$
9	$B = 1100, \bar{B} = 0011$
10	$A = 1110, B = 1101, A \oplus B = 0011$
11	$A = 1110, B = 1100, A\bar{B} = 0010$
12	$A = 1100, B = 1111, \bar{A} + B = 1111$
13	$A = 1100, B = 1100, \overline{A \oplus B} = 1111$
14	$B = 1100$
15	$A = 1110, B = 1111, AB = 1110$
16	One output
17	$A = 1100, B = 1101, A + \bar{B} = 1110$
18	$A = 1100, B = 1111, A + B = 1111$
19	$A = 1111$

Conclusions

In a circuit that has many different types of logic operations to perform, the ALU can be extremely effective. With the proper programming on the function selector inputs, the unit will AND, OR, invert, or combine the inputs in a variety of other ways.

EXPERIMENT 6-7   CHECKING ALU ARITHMETIC

Purpose

To investigate the arithmetic operations of the 74181 ALU.

Parts

Item	Quantity
74181 ALU	1
SPDT switch	8
LEDs	4
Logic probe	1

Procedure

Step 1. The experiment uses the same setup as was shown in Fig. 6-6. Only one change is necessary to put the ALU in the arithmetic mode. The M input (pin 8) is to be grounded.

Step 2. There are two sets of arithmetic operations possible with the 74181 depending on the level of the carry input (pin 7). This experiment will deal with the situation when C_n is high, meaning that there is no carry. The other case will be covered in the discussion which follows, as well as in the explanation of the expected results.

Step 3. The functions performed by the various groupings of pins are the same in the logic or arithmetic mode. You may wish to refresh your memory on their usage by reviewing Table 6-11.

Step 4. A output. Throughout this experiment the words "plus" and "minus" will be used instead of their signs. This procedure avoids any confusion between addition and ORing, which both use the $+$ sign. (The $+$ sign then is used for OR only.) The arithmetic operations that will be verified are in the column of Table 6-14 labeled "no carry."

Input		Expected Output
$A0 = 1$	$S0 = 0$	$F0 = 1$
$A1 = 1$	$S1 = 0$	$F1 = 1$
$B0 = X$	$S2 = 0$	$F2 = 1$
$B1 = X$	$S3 = 0$	$F3 = 1$

Step 5. OR. This operation is identical to the logic function.

Input		Expected Output
$A0 = 0$	$S0 = 1$	$F0 = 1$
$A1 = 0$	$S1 = 0$	$F1 = 0$
$B0 = 0$	$S2 = 0$	$F2 = 1$
$B1 = 1$	$S3 = 0$	$F3 = 1$

The remaining steps are listed in Table 6-15 (page 122). Carry out the operations and check your results.

Step 20. Comparison. Make the A and B outputs equal. Did you find that pin 14 is high? Now change either input. That output then drops to the low level. This comparator output does not depend on any other inputs except the values of A and B.

120

Table 6-14 74181 Arithmetic Functions*

S3	S2	S1	S0	Value of F	
				No Carry (C_n High)	With Carry (C_n Low)
0	0	0	0	A	A plus 1
0	0	0	1	$A + B$	$(A + B)$ plus 1
0	0	1	0	$A + \bar{B}$	$(A + \bar{B})$ plus 1
0	0	1	1	Minus 1 (2's complement)	0
0	1	0	0	A plus $A\,\bar{B}$	A plus $A\,\bar{B}$ plus 1
0	1	0	1	$(A + B)$ plus $A\,\bar{B}$	$(A + B)$ plus $A\,\bar{B}$ plus 1
0	1	1	0	A minus B minus 1	A minus B
0	1	1	1	$A\,\bar{B}$ minus 1	$A\,\bar{B}$
1	0	0	0	A plus AB	A plus AB plus 1
1	0	0	1	A plus B	A plus B plus 1
1	0	1	0	$(A + \bar{B})$ plus AB	$(A + \bar{B})$ plus AB plus 1
1	0	1	1	$A\,B$ minus 1	AB
1	1	0	0	A plus A (double)	A plus A plus 1
1	1	0	1	$(A + B)$ plus A	$(A + B)$ plus A plus 1
1	1	1	0	$(A + \bar{B})$ plus A	$(A + \bar{B})$ plus A plus 1
1	1	1	1	A minus 1	A

*M input low.

Ins and Outs

In the arithmetic mode, the ALU can add, subtract, compare magnitudes, and double the value of the input. There are 12 other lesser-used operations provided as well. As Table 6-15 shows, the outputs vary with the value of C_n. The difference is that the outputs are one greater if C_n is low as compared to those when C_n is high.

The values obtained in the experiments can be shown correct by the same procedure used in the last experiment. Here only selected examples will be proven.

Step	Proof
7	$F = 1111$ (minus 1 in 2's complement)
10	$A = 1111, B = 1100, A - B - 1 = 0010$
12	$A = 1110, B = 1110, A$ plus $B = 1100$ (Ignoring carry out)
16	$A = 1100, A$ plus $A = 1000$ (double)
19	$A = 1111, A - 1 = 1110$

Table 6-15 Experimental Results

Step	A1	A0	B1	B0	S3	S2	S1	S0	F3	F2	F1	F0
			Inputs						Expected Output			
6	0	0	1	0	0	0	1	0	1	1	0	1
7	X	X	X	X	0	0	1	1	1	1	1	1
8	1	0	1	0	0	1	0	0	1	1	1	0
9	1	1	1	1	0	1	0	1	1	1	1	1
10	1	1	0	0	0	1	1	0	0	0	1	0
11	1	1	0	0	0	1	1	1	0	0	1	0
12	1	0	1	0	1	0	0	0	1	1	0	0
13	0	0	0	1	1	0	0	1	1	0	0	1
14	1	1	0	0	1	0	1	0	1	0	1	1
15	1	1	1	1	1	0	1	1	1	1	1	0
16	0	0	X	X	1	1	0	0	1	0	0	0
17	0	1	1	0	1	1	0	1	1	1	0	0
18	0	1	1	0	1	1	1	0	1	0	1	0
19	1	1	X	X	1	1	1	1	1	1	1	0

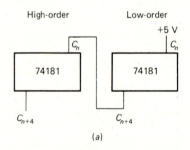

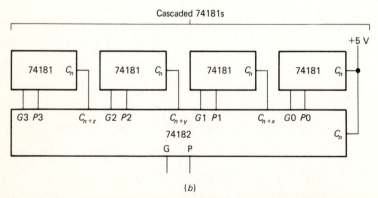

Fig. 6-7 Several configurations of cascaded 74181s can be used, depending on how the carry is to be handled. *(a)* Ripple carry; *(b)* look-ahead carry.

The carry may be handled in various ways. If speed is not of utmost importance, a ripple carry can be used to cascade the ALUs for longer than 4-bit quantities. For that application the ripple carry output (C_{n+4}) becomes the ripple carry input (C_n) for the next-higher stage (see Fig. 6-7). For high-speed operations, the 74182 look-ahead carry generator is used in conjunction with the ALU.

Conclusions

The ALU is a highly versatile package that has logic function and arithmetic operating modes. The arithmetic mode offers addition, subtraction, doubling, magnitude comparison, and other operations. In the logic mode, this device can provide the OR, AND, NOR, NAND, exclusive OR, and comparator outputs, among others. Either ripple carry or look-ahead carry are supported by the outputs available.

EXPERIMENT 6-8   IDENTIFYING PARITY

Purpose

To compute the parity of binary numbers with a parity generator/checker.

Parts

Item	Quantity
74180 parity generator	1
SPDT switch	6
LEDs	2
Logic probe	1

Procedure

Step 1.   Connect the circuit as shown in Fig. 6-8. The upper 4 bits are all high in that circuit, so the *E* through *H* inputs are ones.

Step 2.   Set inputs *A* through *D* to all high. Note the output levels.

Pin	Level
5	*(High)*
6	*(Low)*

Step 3.   Now ground any one of the inputs and repeat your readings.

Pin	Level
5	*(Low)*
6	*(High)*

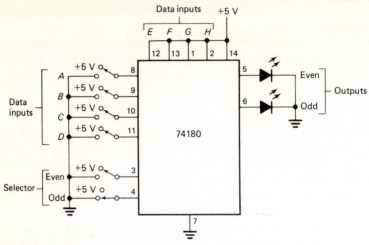

Fig. 6-8 The parity checker/generator provides coding to implement error detection and correction on data input or output.

The *parity* of a binary number is just an indication of whether it has an even or odd count of ones in it. For example, see Table 6-16.

Table 6-16 Parity of Binary Numbers

Binary Number		Quantity of Ones in the Number	Parity
0000	0000	0	Even
0000	0001	1	Odd
0000	0010	1	Odd
0000	1111	4	Even
0100	0101	3	Odd
1111	1111	8	Even

Notice that parity does not depend on the position of the ones, only on the total count.

The 74180 parity generator will tell you if the parity is even or odd. You program it with the settings of pins 3 and 4.

Step 4. Repeat Steps 2 and 3, but first reverse the settings on pins 3 and 4. (Ground pin 3 and tie pin 4 high.) Do you obtain outputs that are just the opposite of those from the previous steps?

Step 5. Now ground a total of any two of the data inputs. This input has six ones, or even parity. With pin 3 high and pin 4 low, what are the outputs?

Pin	Level
5	(*High*)
6	(*Low*)

Step 6. Ground both selector pins 3 and 4. Are both outputs high? How do they change with both selector pins high? You should see both outputs drop to low.

Ins and Outs

The functions for the 74180 are listed in Table 6-17. When the selector inputs are in opposite states, the device supplies an indication of the data parity. If both selectors are in the same state, the outputs are meaningless.

Table 6-17   74180 Parity Generator Functions

Data Parity	Selector Input		Output	
	Even (Pin 3)	Odd (Pin 4)	Even (Pin 5)	Odd (Pin 6)
Even	H	L	H	L
Odd	H	L	L	H
Even	L	H	L	H
Odd	L	H	H	L
X	H	H	L	L
X	H	H	H	H

When the even selector is high and the odd selector is low, the results are indicated by:

Parity of Data	Output Level	
	Pin 5	Pin 6
Even	H	L
Odd	L	H

If the selector input levels are the opposite, the outputs then become:

Parity of Data	Output Level	
	Pin 5	Pin 6
Even	L	H
Odd	H	L

125

Normally the parity of a data word is sent along with it as an extra bit. For example, let the data be $1110\ 0011_2$ and assume that odd parity is to be used. The information transmitted is then

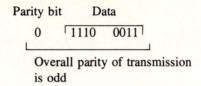

Parity bit   Data

0   1110   0011

Overall parity of transmission
is odd

The most common transmission error is a single bit changing state (that is, one becomes zero or vice versa). Using the first example, let the LSB be changed in transmission. The received information is

Overall parity is even

0   1110   0010
                         ↑
                    Bit with error

Because the received information has even parity, the error in transmission is easily detected and the data can be sent again.

Conclusions

The 74180 can generate or check the parity of 8 bits. The parity indicates whether there is an even or odd count of ones in the data. The major use of parity is to detect errors that occur when digital quantities are moved from one location to another.

7

Flip-Flops

The circuits used in the experiments to this point can all be characterized by their direct reliance of the output on the combination of input signals. When any of the inputs are changed or removed, the outputs are altered. Next you will learn about another class of circuits which differ markedly in how the outputs react. Called *sequential circuits*, these devices have a memory. The output depends not only on the present input, but also on the previous state of the output. Just as gates are the foundation of the combinatorial circuits, flip-flops form the basis of sequential circuitry. This series of experiments will demonstrate how the simplest flip-flops are built and will then investigate the different types of flip-flops.

EXPERIMENT 7-1   MAKING AN *R-S* FLIP-FLOP

Purpose

To build a simple *R-S* flip-flop.

Parts

Item	Quantity
7400 quad NAND	1
7405 hex inverter	1
SPDT switch	2
LEDs	2

Procedure

Step 1.   Connect the circuit as shown in Fig. 7-1.

Step 2.   With the switches, apply a high level to the S input and a low to the R input. What is the state of the output? *(Q is high and $\bar{Q}$ is low)* As this step shows, the flip-flop is an obviously different type of circuit than a gate. It has two outputs instead of one. The outputs are labeled Q and $\bar{Q}$ because they are always in opposite states *under normal operating conditions*. When the Q output is high and $\bar{Q}$ low, the flip-flop is said to be *set*. In the alternate state, that is, $\bar{Q}$ high and Q low, the device is *reset*. Now you see why this device is named an *R-S flip-flop*.

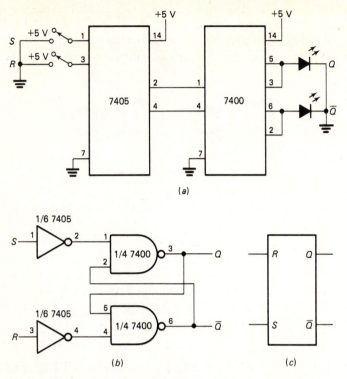

Fig. 7-1 The *R-S* flip-flop can latch the last input value. In this way the device acts as a single-bit memory. *(a)* Schematic; *(b)* logic diagram; *(c)* symbol.

Step 3. Now ground both inputs. How are the outputs affected? *(No change)* This experiment shows another important feature of a flip-flop. It *latches* the output state, so the inputs can be removed without changing the outputs. These levels remain until another change is made in the *R-S* inputs.

Step 4. Change the settings so that *R* is high and *S* is low. Do you observe that *Q* goes low and $\overline{Q}$ becomes high? This step shows how the latch is reset.

Step 5. Now switch both inputs low. Again the outputs should be latched.

Step 6. Switch the *S* and *R* inputs to the high level. Try this several times and examine how the outputs react to inputs changing to the high state simultaneously. Also try switching them from low to high in succession. Because the outputs depend on what event occurred previously (and not just on the state of the inputs), this situation is called the *indeterminate,* or sometimes the forbidden, input combination. Save this circuit. It will be used in the next experiment.

Ins and Outs

The *R-S* flip-flop can be described by a state table (see Table 7-1). As usual, a one means a high level and a zero a low level. Table 7-1 uses a unique way to indicate that

128

simultaneous zero inputs on R and S cause no change. The Q_0 and $\bar{Q}_0$ symbols refer to the *original* values of the outputs, so Table 7-1 is saying that the final values of the outputs are the same as the starting values before the zero inputs. As you saw in the experiments, just knowing that both inputs are high is not sufficient to determine the output states.

Table 7-1   *R-S* Flip-Flop State Table

Input		Output	
R	S	Q	$\bar{Q}$
0	1	1	0
1	0	0	1
0	0	Q_0	$\bar{Q}_0$
1	1	Indeterminate	

Conclusions

The *R-S* flip-flop latches the last settings of the inputs. If S is high and R is low, the device is set. On the other hand, a high R input and a low S input resets the flip-flop. Because the outputs are not predictable from the inputs, two highs into the latch result in an undetermined output.

EXPERIMENT 7-2   CONVERTING THE LATCH TO A CLOCKED *R-S* FLIP-FLOP

Purpose

To investigate clocked flip-flops.

Parts

Item	Quantity
7400 quad NAND	1
SPDT switch	3
LEDs	2

Procedure

Step 1.   Wire the circuit as shown in the schematic of Fig. 7-2.

Step 2.   With the clock input (Ck) connected to 5 V, test the effect that various inputs have on the Q and $\bar{Q}$ outputs. Record your answers in Table 7-2.

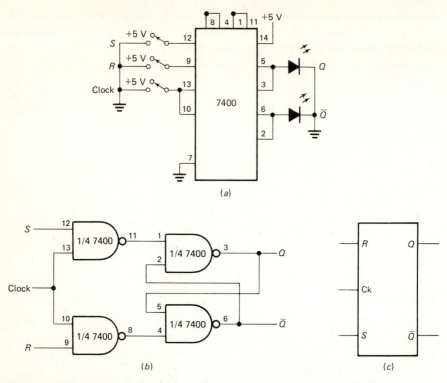

Fig. 7-2 The clocked *R-S* flip-flop synchronizes the changing states of the outputs with the clock input. Only when the clock is high can the outputs change. *(a)* Schematic; *(b)* logic diagram; *(c)* symbol.

Table 7-2

		Your Results		Expected Results	
R	S	Q	$\bar{Q}$	Q	$\bar{Q}$
0	1			1	0
1	0			0	1
0	0			Q_o	$\bar{Q}_o$
1	1			Indeterminate	

Step 3. With the *R* input low and the *S* input high, ground the clock. Then change the inputs. What is the result? *(No change in outputs)*

Step 4. Repeat Step 3 with initial conditions of *S* high and *R* low. *(Again, no change)*

Ins and Outs

The clocked *R-S* flip-flop operates in just the same way as does the circuit in Experiment 7-1 when the clock is high, thus *enabling* the flip-flop. When the clock is low, no change in the outputs can occur. In effect, the leftmost NAND gates in Fig. 7-2b act to disable/enable the flip-flop shown on the right-hand side of the diagram. Circuits that are clocked are called *synchronous* circuits. If no clock is used, the circuit is *asynchronous*.

The functions of the clocked *R-S* flip-flop are summarized in Table 7-3. The square-wave symbol in the clock column means that a high-level pulse is to be used. During the time the clock is high, data inputs to the device must be held constant. After the clock falls, the *R* and *S* inputs can be changed at will.

Table 7-3   Clock *R-S* Flip-Flop Functions

Input			Output	
Clock	R	S	Q	$\bar{Q}$
L	X	X	Q_o	$\bar{Q}_o$
⊓	0	1	1	0
⊓	1	0	0	1
⊓	0	0	Q_o	$\bar{Q}_o$
⊓	1	1	Indeterminate	

Conclusion

The synchronous *R-S* flip-flop uses a clock to enable the latch after the data inputs are at the correct levels. Data inputs are only changed during the interval when the clock pulse is low.

EXPERIMENT 7-3   REVIEWING *R-S* FLIP-FLOPS

Purpose

To obtain practice in the use of integrated-circuit flip-flops.

Parts

Item	Quantity
74L71 flip-flop	1
SPDT switch	5
LEDs	2

Procedure

Step 1. Wire the circuit as shown in Fig. 7-3.

Step 2. With the preset and clear inputs both high, verify that the data inputs produce the correct outputs. Remember to switch the clock high after the correct *R* and *S* settings are made. Then switch it low when you change the inputs.

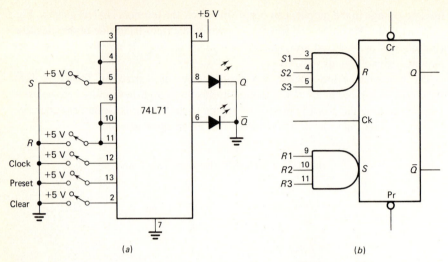

Fig. 7-3 The 74L71 is an *R-S* flip-flop that is provided with AND gate inputs as well as preset and clear selections. *(a)* Schematic; *(b)* logic diagram.

Step 3. Ground the preset input and change the clock and data inputs in various combinations. What are the results? *(Q high and $\bar{Q}$ low)*

Step 4. Switch the preset input to 5 V and ground the clear input. Again change the other inputs. What do you observe? *(Q low and $\bar{Q}$ high)*

Step 5. Ground both the preset and clear inputs. What happens? *(Both Q and $\bar{Q}$ high)* What happens when you switch the preset and clear back to high levels? *(Indeterminate results)*

Ins and Outs

The 74L71 is best understood from the function table provided in Table 7-4. The preset or clear inputs will override all others. These two inputs are used to establish an initial and known state of the flip-flop prior to some sequence of operations. Most often a flip-flop is cleared; however, it is sometimes better to start in a preset state. So both types of inputs are provided. After initialization, these two inputs are left at the high level.

Another feature to notice is that a clock pulse activates the data inputs, thus allowing them to change the outputs. If a clock pulse is involved, the flip-flop is said to be *level-triggered.* Another type of timing, *edge-triggering,* is investigated in a later experiment.

Conclusions

The 74L71 forms a clocked, *R-S* flip-flop. In addition to the data inputs, preset and clear pins are also provided. These last two inputs are used to set up starting conditions, but they are not normally used during other times.

Table 7-4 74L71 Functions

Input					Output	
Preset	Clear	Clock	S	R	Q	$\bar{Q}$
L	H	X	X	X	1	0
H	L	X	X	X	0	1
L	L	X	X	X	1*	1*
H	H	⌐_⌐_	0	0	Q_o	$\bar{Q}_o$
H	H	⌐_⌐_	1	0	1	0
H	H	⌐_⌐_	0	1	0	1
H	H	⌐_⌐_	1	1	Indeterminate	

*Not a stable condition (see explanation in text).

The 74L71 provides AND gates for the R and S inputs, as Fig. 7-3b shows. These AND gates each fan in three separate inputs, thus combining them to form the inputs to the flip-flop. If only a single S and R input is needed, the other inputs are simply connected together, as you did in this experiment.

EXPERIMENT 7-4 COMPARING *J-K* AND *R-S* FLIP-FLOPS

Purpose

To study a dual *J-K* flip-flop.

Parts

Item	Quantity
7476 dual *J-K* flip-flop	1
SPDT switch	5
LEDs	2

Procedure

Step 1. Construct the circuit as shown in Fig. 7-4.

Step 2. The preset and clear inputs on this flip-flop work in the same manner as those on the *R-S* flip-flop of the previous experiment. By use of the switches, verify this operation.

Step 3. Connect the preset and clear inputs to 5 V.

Step 4. This flip-flop also uses level triggering, so the clock should be switched to the high level after each change of inputs. By then switching the clock low, the outputs will not change until the next clock pulse. Follow this clocking scheme for the remainder of the steps in this experiment.

Step 5. Show that a high on the J input and low on the K input sets the Q output and clears the $\bar{Q}$ output.

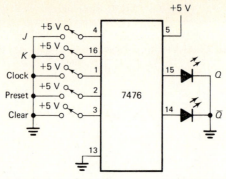

Fig. 7-4 The 7476 is a dual *J-K* flip-flop package. Two complete *J-K* flip-flops with their preset and clear inputs are included in one DIP.

Step 6. Similarly examine the outcome of a low *J* and high *K* input. *(Q low and $\bar{Q}$ high)*

Step 7. What happens if both *J* and *K* become low? *(No change in previous output state)*

Step 8. Establish the outputs in any convenient state. Then switch both *J* and *K* to the high level. Repeat with other initial conditions for the outputs. What do you find? *(The Q and $\bar{Q}$ outputs both change to the opposite states.)*

Ins and Outs

In many respects, the *J-K* flip-flop resembles the *R-S* flip-flop. When *J* and *K* are at opposite voltage levels, the outputs change accordingly. Low levels on both data inputs cause the outputs to retain the previous setting. It is when the *J* and *K* inputs are both high that a difference is noted from that of the *R-S* flip-flop.

This input combination causes the *Q* and $\bar{Q}$ outputs to flip to their complements. The simultaneous change in the outputs in this fashion is often referred to as *toggling*. Table 7-5 lists the functions of the *J-K* flip-flop.

Table 7-5   7476 *J-K* Flip-Flop Functions

Input					Output	
Preset	Clear	Clock	*J*	*K*	*Q*	*$\bar{Q}$*
L	H	X	X	X	1	0
H	L	X	X	X	0	1
L	L	X	X	X	1*	1*
H	H	⎍	0	0	Q_o	$\bar{Q}_o$
H	H	⎍	1	0	1	0
H	H	⎍	0	1	0	1
H	H	⎍	1	1	Toggle	

*Unstable output condition.

Conclusions

The 7476 *J-K* flip-flop can be set or cleared with the preset or clear inputs. Its outputs respond to the *J* and *K* inputs when the clock is high. In the situation when both data inputs are ones, that is, high, the outputs toggle.

EXPERIMENT 7-5   EXAMINING *D* FLIP-FLOPS

Purpose

To learn how to use *D* flip-flops.

Parts

Item	Quantity
7474 dual *D* flip-flop	1
SPDT switch	4
LEDs	2

Procedure

Step 1.   Construct the circuit as shown in Fig. 7-5*a*.

Step 2.   Test the preset and clear inputs (pins 1 and 4). Do they operate as you expected?

Step 3.   Switch the clock input high.

Step 4.   Does changing the *D* input change the outputs? *(No)*

Step 5.   Set the *D* input high. Now switch the clock low then back to high. Record your results. *(Q high, $\bar{Q}$ low)*

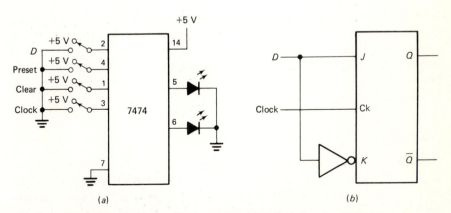

Fig. 7-5   The *D* flip-flop delays the input signal briefly before it appears at the *Q* terminal. The *J-K* flip-flop forms the basis of the *D* flip-flop. *(a)* Schematic; *(b)* logic diagram.

135

Step 6. Ground the D input. Switch the clock between ground and 5 V again. Did the outputs change? *(Yes, Q became low and $\bar{Q}$ high)*

Ins and Outs

The D, or delay, flip-flop causes the Q output to follow the D input after the propagation delay of the circuit. In addition to this IC being a D flip-flop, it also exhibits *edge-triggering*. It was this type of clocking that prevented the outputs from changing in Steps 5 and 6 even though the clock was initially high. Only if the clock level was dropped and raised did the outputs change. The *rising* edge of the clock transitioning from low to high was the determining factor. (Other types of edge-triggered flip-flops may react to the *falling* edge of the clock.)

The functions table for this flip-flop (Table 7-6) indicates the rising-edge-triggering by an arrow pointing upward. Falling-edge-triggering is signified by a downward-pointing arrow.

Table 7-6   7474 D Flip-Flop Functions

Preset	Clear	Clock	D	Q	$\bar{Q}$
_____ Input _____				Output	
L	H	X	X	1	0
H	L	X	X	0	1
L	L	X	X	1*	1*
H	H	↑	1	1	0
H	H	↑	0	0	1
H	H	L	X	Q_o	$\bar{Q}_o$

*Unstable output condition.

The D flip-flop can be constructed from a J-K flip-flop and an inverter as shown in Fig. 7-5b. Of course, it is more convenient to work with the prepackaged 7474 or other types of D flip-flops. This logic diagram does make the D flip-flop operation clear, though. When the J (also labeled D) input is high, the K input is forced low. The outputs are Q high and $\bar{Q}$ low. If D were low, J would be low also, but K would be high. Then Q would fall and $\bar{Q}$ would rise.

Conclusions

The D flip-flop Q output follows the level of the input. Because the 7474 flip-flop is rising-edge-triggered, changes occur only during a clock transition from low to high level. A falling-edge-triggered flip-flop changes when the clock goes from high to low level.

EXPERIMENT 7-6   CONSTRUCTING A *T* FLIP-FLOP

Purpose

To make a *T* flip-flop from a *J-K* flip-flop.

Parts

Item	Quantity
7476 dual *J-K* flip-flop	1
SPDT switch	3
LEDs	2

Procedure

Step 1.   Wire the flip-flop as shown in Fig. 7-6.

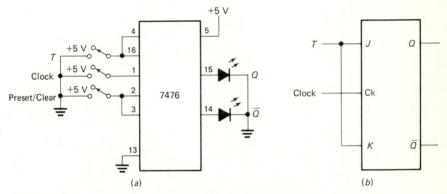

Fig. 7-6   By tying the data inputs together, the *J-K* flip-flop becomes a *T*, or toggle, flip-flop. *(a)* **Schematic;** *(b)* **logic diagram.**

Step 2.   You have already verified the operation of the clock, preset, and clear inputs to this device. Before proceeding with the experiment, you may wish to review those results.

Step 3.   Switch the clock high.

Step 4.   Input a zero (ground) on the *T* input (pins 4 and 16). Do the outputs change? *(No)*

Step 5.   Switch the *T* input high. How do the outputs react? *(Toggle)*

Step 6.   Switch the *T* input between ground and 5 V several times. What happens with each occurrence? *(Toggle)*

Ins and Outs

By now you must have realized that the *T* flip-flop is named for toggle. With each new high input, the outputs are toggled. An input of a low level does not change the outputs. Table 7-7 is a brief summary of *T* flip-flop operation. Because the *T* flip-flop is so simply constructed from a *J-K* flip-flop, no one manufacturers a toggle flip-flop.

Table 7-7   *T* Flip-Flop Functions

T	Output	
	Q	$\overline{Q}$
1	Toggle	
0	Q_o	$\overline{Q}_o$

Conclusion

The *T* flip-flop alters its output states on each high input. A *J-K* flip-flop is converted to a *T* flip-flop if the two data inputs are connected.

EXPERIMENT 7-7   DEBOUNCING A SWITCH

Purpose

To show how a simple flip-flop eliminates transients in mechanical switches.

Parts

Item	Quantity
7400 quad NAND	1
330-Ω resistor	2
Push-button switch	1
Decade counter (see Chap. 1)	1

Procedure

Step 1.   Wire the counter as shown in Chap. 1. The input of the counter is attached to the switch (see Fig. 7-7).

Step 2.   Open and close the switch. The decade counter increments by one each time the switch makes contact. Do you find that on closing the switch the counter jumps by some random amount each time?

Step 3.   Using the 7400 build the circuit in Fig. 7-8. Attach the output to the decade counter.

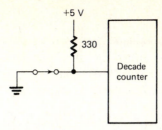

Fig. 7-7 The decade counter shows that a switch makes and breaks contact many times.

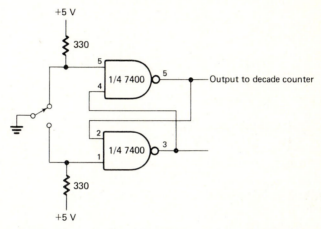

Fig. 7-8 The latch debounces the switch, thus guaranteeing positive contact every time.

Step 4. Now close the switch. Your counter should increment by one on each attempt.

Ins and Outs

The latch *debounces* the mechanical switch that makes and breaks contact many times before the transient effects die out. This circuit is often used in connecting a ''reset'' or ''clear'' switch to digital equipment. If the switch is not debounced, the equipment mistakenly reacts to the numerous false openings and closures of the switch.

Conclusions

Mechanical switches must be debounced prior to connecting them into logic circuits. If this conditioning of the switch input is not provided, the digital equipment may receive several false make and break signals each time the switch is closed. The *R-S* latch constructed from a 7400 circuit is an economical way of supplying the debouncing.

8

Registers and Counters

By coupling 1-bit flip-flops together, you can build storage devices that can retain the values of many bits. These configurations, or registers, are used not only to temporarily store multibit values, but also are frequently applied in shifting the bits as a unit. Another application of flip-flops is in counters. These circuits provide an indication of the total number of input pulses that have been received. Counters additionally serve to reduce high-frequency clocks to lower-frequency signals. Most digital watches and clocks have counters which reduce the line frequency or an oscillator frequency to the number of seconds, minutes, and hours that have elapsed.

EXPERIMENT 8-1   CONTROLLING A SHIFT REGISTER

Purpose

To determine the outputs of a shift register with multiple shifting modes.

Parts

Item	Quantity
74194 shift register	1
LEDs	4
SPDT switch	9
Clock generator (see Chap. 1)	1

Procedure

Step 1.   The circuit is shown in Fig. 8-1. Do not connect the clock generator at this time.

140

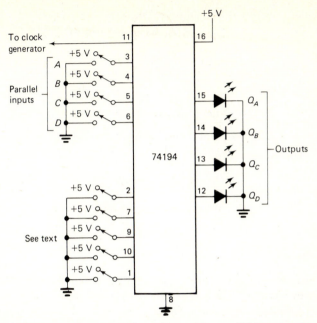

Fig. 8-1 The 74194 is a universal shift register with parallel input and output. It can shift left or right on each clock pulse.

Step 2. The first action that we will take is to clear the outputs. Grounding pin 1 will force all outputs to zero. After clearing the register, attach pin 1 to 5 V.

Step 3. This register is rising-edge-triggered. As long as the clock remains low, the outputs do not change. In the following steps, the clock generator should be switched on after all input conditions are established to observe how the outputs respond.

Step 4. Pins 9 and 10 program the mode of the 74194. Set both of these inputs to the high level. Set pins 2 and 7 low.

Step 5. Set the *A*, *B*, *C*, and *D* parallel input switches as follows:

> *A* high
>
> *B* low
>
> *C* high
>
> *D* low

Step 6. Turn on the clock. What outputs develop? (*Each output agrees with its corresponding inputs:* $Q_A = A$, $Q_B = B$, $Q_C = C$, and $Q_D = D$)

Step 7. Switch the clock off. Switch the shift right control (pin 2) to the high level. Ground pin 9. Manually switch the clock input (pin 11) low and then high. (Do not use the clock generator.) How are the output levels changed? ($Q_A = high$, $Q_B = high$, $Q_C = low$, $Q_D = high$)

Step 8. Return the shift right control (pin 2) to ground. Manually provide a low to high clock transition as in Step 6. What do the outputs do? $(Q_A = low, Q_B = high, Q_C = high, Q_D = low)$

Step 9. Set pin 10 high and pin 9 low.

Step 10. Set the left shift (pin 7) to the high level. On a manual clock pulse, how do the outputs change? $(Q_A = high, Q_B = high, Q_C = low, Q_D = high)$

Step 11. Switch pin 7 low. What happens on the next manual switching of the clock from low to high? $(Q_A = high, Q_B = low, Q_C = high, Q_D = low)$

Step 12. Reconnect the clock generator to pin 11. Switch pin 7 high. Do you observe the output pattern shifting left? Reverse the settings of pins 9 and 10, keeping one high and the other low. Does the pattern reverse direction as you reverse the levels on pins 9 and 10?

Ins and Outs

The 74194, as used in this experiment, is a parallel input, parallel output (PIPO) shift register. Other shift registers may operate as parallel input, serial output (PISO), serial input, parallel output (SIPO), or serial input, serial output (SISO). The mode control inputs, $S0$ and $S1$, determine the operation performed. These mode inputs (pins 9 and 10) must be high for parallel loading. Table 8-1 shows how their settings affect right-shifting or left-shifting.

Table 8-1 74194 Shift Register Functions

	Inputs									Outputs			
	Mode			Serial		Data							
Clear	$S0$	$S1$	Clock	Left	Right	A	B	C	D	Q_A	Q_B	Q_C	Q_D
L	X	X	X	X	X	X	X	X	X	L	L	L	L
H	X	X	L	X	X	X	X	X	X	Q_{A0}	Q_{B0}	Q_{C0}	Q_{D0}
H	H	H	↑	X	X	Any values				A	B	C	D
H	L	H	↑	X	H	X	X	X	X	H	Q_{An}	Q_{Bn}	Q_{Cn}
H	L	H	↑	X	L	X	X	X	X	L	Q_{An}	Q_{Bn}	Q_{Cn}
H	H	L	↑	H	X	X	X	X	X	Q_{Bn}	Q_{Cn}	Q_{Dn}	H
H	H	L	↑	L	X	X	X	X	X	Q_{Bn}	Q_{Cn}	Q_{Dn}	L
H	L	L	X	X	X	X - X	X	X	X	Q_{A0}	Q_{B0}	Q_{C0}	Q_{D0}

A right shift moves the previous output of Q_D to Q_C, Q_C to Q_B, and so on. Suppose the original levels were

$$Q_A \quad Q_B \quad Q_C \quad Q_D$$
$$H \quad L \quad H \quad L$$

After the shift, the pattern becomes

$$Q_A \quad Q_B \quad Q_C \quad Q_D$$
$$? \quad\ H \quad\ L \quad\ H$$

What about the new level of Q_A? No bit is available to move into that position. The answer in the case of the 74194 depends on the level of the shift right input (pin 2). That input level becomes the next value for Q_A. To write the description more compactly, a right shift may be described as

$$Q_A = \text{shift right input level}$$

$$Q_B = Q_{An}$$

$$Q_C = Q_{Bn}$$

$$Q_D = Q_{Cn}$$

Where the subscript n represents the level of that particular output before the last clock transition.

For a left shift, therefore, the levels become

$$Q_A = Q_{Bn}$$

$$Q_B = Q_{Cn}$$

$$Q_C = Q_{Dn}$$

$$Q_D = \text{shift left input level}$$

Conclusions

The 74194 shift register operates as a parallel input register and as a right- or left-shift register. The operation is controlled by the settings of the mode selector switches. A separate clear input forces all outputs to the low level.

EXPERIMENT 8-2 SHOWING SERIAL SHIFT REGISTER OPERATIONS

Purpose

To connect a serial shift register circuit.

Parts

Item	Quantity
7491 shift register	1
SPDT switch	2
LEDs	2
Clock generator	1

Procedure

Step 1. Wire the circuit as shown in Fig. 8-2.

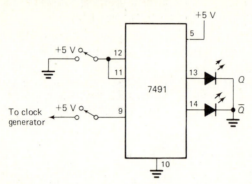

Fig. 8-2 **The 8-bit serial shift register stores an 8-bit quantity and makes it available 1 bit at a time. The two inputs to the register are ANDed before being stored.**

Step 2. Connect the input (pins 11 and 12) to 5 V.

Step 3. Switch on the clock generator and observe the output. *(After a short delay, the Q output becomes high, and $\bar{Q}$ low)*

Step 4. Now ground the input. What happens? *(After a slight delay, Q goes low and $\bar{Q}$ high)*

Step 5. Attempt to measure the time delay with a watch with a second hand. As the input level is reversed, note the time and record the interval before the output changes. You should try this step a few times and average your results. *(About 8 s)*

Ins and Outs

The 7491 is an SISO shift register that can hold a total of 8 bits. The register is composed of *R-S* flip-flops with the necessary input gates and a synchronizing clock signal. On each clock pulse, the register contents shift 1 bit toward the output. That effect caused the time delay in Step 5.

The two input pins were connected together in this experiment. If the inputs are used individually, the value placed into the first bit position is equal to *A* AND *B* (pin 12 is the *A* input and pin 11 is *B*). The action of the register is summarized in Table 8-2.

Table 8-2 7491 Shift Register Functions

Input		Output	
A	*B*	*Q*	*$\bar{Q}$*
1	1	1	0
0	X	0	1
X	0	0	1

*After eight clock pulses.

Conclusions

This register can store 8 bits of data and clock them out one at a time. Temporarily holding an 8-bit quantity, also called a *byte,* is frequently necessary in digital computers, control systems, and communication devices.

EXPERIMENT 8-3 CALCULATING WITH A DECADE COUNTER

Purpose

To examine a count-by-10 circuit.

Parts

Item	Quantity
7400 quad NAND	1
7490 decade counter	1
330-Ω resistors	6
LEDs	4
Push-button switch	1

Procedure

Step 1. As the schematic in Fig. 8-3 shows, this circuit makes use of the switch debouncer from Chap. 7. Construct this circuit.

Step 2. The decade counter increments the display on each switch closure. The

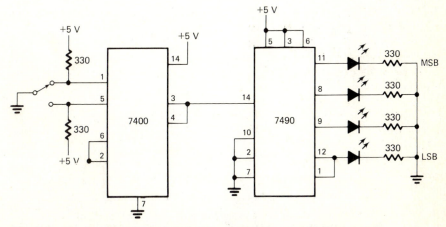

Fig. 8-3 The decade counter is able to sequentially count 10 input pulses or to divide by 10 with a slight change in the wiring.

LEDs are read in binary. That is, the LED on pins 1 and 12 represents the LSB and the one on pin 11, the MSB. A pattern of

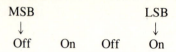

MSB			LSB
↓			↓
Off	On	Off	On

is read 0101_2 or 5 in decimal.

Step 3. Starting with a zero count, open and close the switch and observe the sequence of the LEDs. *(Counts from 0 to 9)*

Step 4. Disconnect pins 1 and 14 of the counter.

Step 5. Connect the output of the 7400 to pin 1 of the 7490.

Step 6. Connect pin 14 to pin 11 of the counter.

Step 7. Count a series of switch actions, 25 or more. How does the LED on pin 12 react? *(Lighted on each tenth input)*

Ins and Outs

The original configuration of the counter permits it to perform a BCD count from zero up to nine. Another function of a counter is to act as a divider. By changing the wiring slightly, the counter became a *divide-by-10* circuit. Such dividers are widely used to change a high-frequency input to a more usable lower frequency.

An excellent example of a divider application is the digital clock. A simplified block diagram of such a clock is shown in Fig. 8-4. The line frequency of 60 Hz is divided by 60 to convert the signal to one pulse per second. (The dividers here do not divide by 10, but by 60. The principle of operation of these dividers is similar to the divide-by-10 you used.) By dividing the number of seconds by 60, the minutes are obtained. One final division gives the hours. Obviously missing from this diagram is the essential circuitry necessary in order to set the clock at the correct starting time, recognize noon or midnight, or sound the alarm.

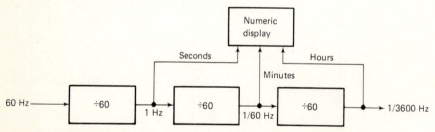

Fig. 8-4   A digital clock can be pictured as a series of dividers. The output of each divider becomes the number of elapsed seconds, minutes, and hours.

Conclusions

A counter increments its output, which can be shown as BCD, by one on each new input. With minor changes, a counter becomes a divider that can divide the input frequency by a fixed amount.

EXPERIMENT 8-4   TESTING A BINARY COUNTER

Purpose

To become familiar with binary counters.

Parts

Item	Quantity
7476 dual *J-K* flip-flop	2
7411 triple AND	1
LED	1
SPDT switch	1
Logic probe	1

Procedure

Step 1.   Connect the circuit as shown in Fig. 8-5*a*. Compare the logic diagram in Fig. 8-5*b* with the schematic to understand the flow through the circuit. Note that the *J-K* flip-flops are connected as *T* flip-flops.

Step 2.   The *T* input to the first flip-flop is always 1. We guarantee this value by wiring it to the power supply. Verify this level with a logic probe.

Step 3.   All inputs to the circuit are made by clock pulses. You create the pulse by throwing the SPDT switch to 5 V and then to ground. With the logic probe on pin 15 of the 7476 flip-flop number 1, perform this switching several times. What happens? *(The output becomes high on alternate inputs)*

Step 4.   Continue throwing the switch until the LED on the output glows. This may require from zero to eight switch actions.

Step 5.   Now input eight clock pulses. Describe what happens to the output LED. *(It goes off after the first input pulse and stays off until the eighth pulse)*

Step 6.   Starting with the output LED on, attach the logic probe to pin 11 of 7476 number 1. Count how many inputs occur between this output switching from the high to low level. *(2)*

Step 7.   Repeat Step 6, but use pin 15 of 7476 number 2. How many inputs are needed for the change? *(4)*

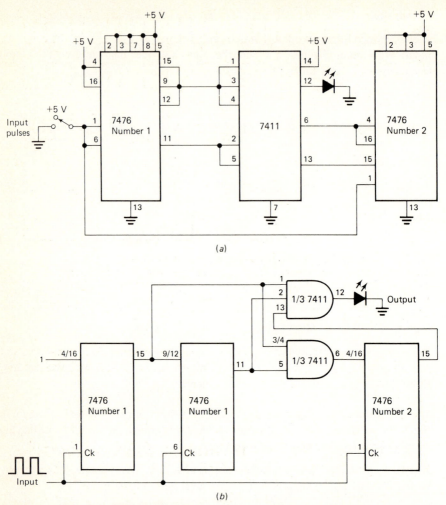

Fig. 8-5 A binary counter always has a maximum count that is a power of 2. In this case the counter provides eight states. *(a)* Schematic; *(b)* logic diagram.

Ins and Outs

The binary counter will reach a maximum which depends on the number of flip-flops. If there are n flip-flops, the count will range from zero to $2^n - 1$. The operation of this counter can readily be seen on a timing diagram, like the one in Fig. 8-6.

As you can see, the Q output of the first flip-flop changes state after every input pulse, the second flip-flop on every other input, and the third flip-flop on every fourth input. The circuit output is high for the length of one input pulse after eight inputs have been received. Because of the sequence of events, this circuit is sometimes called a *ripple counter*.

148

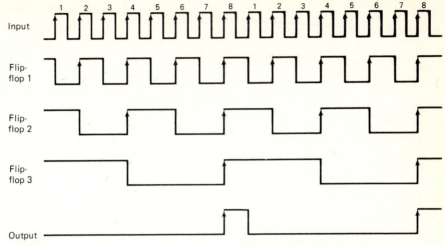

Fig. 8-6 A timing diagram for a binary counter shows that each succeeding stage requires twice the number of inputs as the preceeding one to change states.

Conclusions

The binary counter is based on a power-of-2 counting sequence. Each flip-flop stage doubles the count of the previous stage. A timing diagram is useful in understanding how this circuit maintains its counting sequence.

EXPERIMENT 8-5   CONTROLLING SYNCHRONOUS 4-BIT COUNTERS

Purpose

To review the operation of counters and to examine preset operations.

Parts

Item	Quantity
74163 counter	1
LEDs	4
SPDT switch	8

Procedure

Step 1.   Wire the circuit as shown in Fig. 8-7.

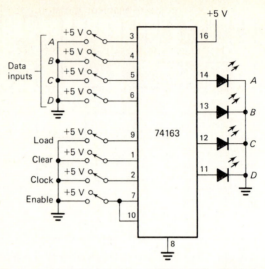

Fig. 8-7 The synchronous counter provides a preset option to begin counting at an arbitrary value. This counter accepts hexadecimal inputs and operates as a hexadecimal counter. A similar decade counter is available as the 74160/74162.

Step 2. Set the switches as follows:

Data Inputs	Don't Care
Clock	H
Load	H
Enable	L
Clear	H

Step 3. Switch the clear low, input a clock pulse, and then switch clear back to high. All the output LEDs should go off.

Step 4. Set the data input switches as follows:

A	L
B	L
C	H
D	H

This input represents a hexadecimal value of C_{16} (or 12_{10}).

Step 5. Switch the load input low, input a clock pulse, and then switch the load input high. How do the outputs change? (*A and B low, C and D high*)

Step 6. Try to input several clock pulses. Do the outputs change? (*No*)

Step 7. The counter has not yet been enabled; therefore, the outputs did not respond. Connect the enable switch to 5 V.

Step 8. Now input a clock pulse. Do the outputs change? *(Yes)*

Step 9. Table 8-3 shows the present count and the outputs for several following values. Verify the table entries. Note that after 16 entries, the counter has returned to the original value.

Table 8-3   Counter Outputs

Clock Input Pulse	Output			
	D	C	B	A
1	H	H	L	H
2	H	H	H	L
3	H	H	H	H
4	L	L	L	L
5	L	L	L	H
6	L	L	H	L
7	L	L	H	H
8	L	H	L	L
9	L	H	L	H
10	L	H	H	L
11	L	H	H	H
12	H	L	L	L
13	H	L	L	H
14	H	L	H	L
15	H	L	H	H
16	H	H	L	L
Return to starting point				

Step 10. Test the counter from other originating points. You may want to start from zero, 8_{10}, or 15_{10}, for example. Test the return to the original value after 16 inputs.

Ins and Outs

This synchronous counter allows the user to preset the values of the outputs at start. The clear function is synchronous also, so the outputs become low with a low level on the clear input after the next clock pulse.

The preset values are accepted when the load signal is pulsed low. To react to input pulses, the counter must first be enabled. Figure 8-8 is a timing diagram for this counter.

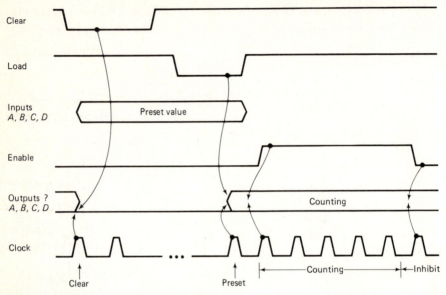

Fig. 8-8   Timing diagram for the 74163 counter.

Conclusions

The 74163 counter is much more sophisticated than those of earlier experiments in this chapter. Its control supports a variety of clear, preset, and enable functions. The sequence of these operations significantly affects the output number. Clear and load are synchronous, meaning that the action is not carried out until the next clock pulse.

EXPERIMENT 8-6   USING AN UP/DOWN COUNTER

Purpose

To examine a counter that increments or decrements.

Parts

Item	Quantity
74192 counter	1
SPDT switch	4
LEDs	4

Procedure

Step 1. Wire the circuit as shown in Fig. 8-9. This counter, like the 74163, can be preset with data inputs. To simplify the setup, these inputs will be connected to ground. As a variation of the basic experiment, you may use switches on these inputs just as in the previous experiment.

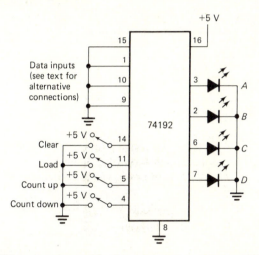

Fig. 8-9 The 74192 counter is capable of incrementing or decrementing the output value. If the countup is used, the counter increments, but if the countdown is selected, it decrements.

Step 2. Set the switches as follows:

Clear	L
Load	H
Count up	H
Count down	H

Step 3. Switch the clear high and then back to low. The outputs should all clear.

Step 4. Set load low and then high. The action initializes the outputs at zero.

Step 5. Pulse the count-up input. On each pulse the output will increment by 1.

Step 6. What happens after the tenth input pulse? (*This is a decade counter, so after an output of 9, the displays return to zero*)

Step 7. Count up to some nonzero value. Switch the count-up input high. Now pulse the count-down input. What are the results? (*The outputs decrement to 0, then 9, 8, 7, and so on*)

153

Ins and Outs

This counter differs from the 74163 in that it (1) is not synchronous, as a clock input is not used, (2) is a decade (BCD) counter instead of a hexadecimal counter, and (3) will increment or decrement. The timing diagram for this counter is shown in Fig. 8-10. The typical operating sequence is clear, load, then count either up or down.

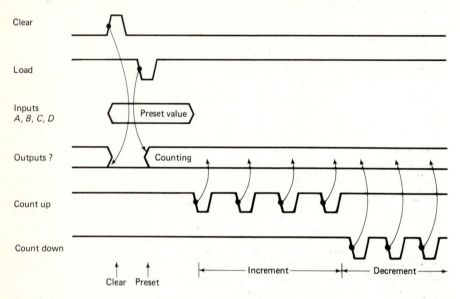

Fig. 8-10 A timing diagram for typical 74192 counting actions.

Conclusions

The 74192 is an advanced counter that satisfies all normal requirements as well as being able to decrease the count. The clear and load operations do not require a clock input, so this is an asynchronous counter.

154

9

Memories

The values that are processed in digital equipment must be stored for later use or to save answers. Memories serve the purpose of allowing storage and retrieval of data. There are several types of memories in use today. Perhaps most common are the integrated-circuit memories, which are available in two basic forms.

Memories that permit numbers to be stored and later read back are called *read/write* or *random-access memories* (RAM). Another useful type of memory does not allow its contents to be changed. *Read-only memory* (ROM) can only be read; its contents are unchangeable. There are a group of memories that fall into the division between ROMs and RAMs. Their contents remain fixed for long periods of time but can be changed with proper equipment. These devices are programmable read-only memories (PROMs). In addition to the TTL and MOS based memories, new technologies are developing *charge-coupled devices* (CCDs) and *bubble memories*.

MEMORY SPECIFICATIONS

A memory may be thought of as a collection of mailboxes, as at the post office. Each box has an address, and information can be placed into any box. The address for each memory location is a number, and the *size* of the memory determines how many addresses there can be. For example, a memory that has 256 words has an address range of 0_{10} to 255_{10}. (The first address is always zero, and the last is one less than the capacity.) For larger memories, a shorter notation is used to indicate the capacity. The letter K stands for multiples of 1024, such as 4K, which indicates a capacity of

$$4 \times 1024 = 4096 \text{ addressable locations}$$

Each memory location holds a certain number of bits often called a *memory word*. If the locations each hold 2 bits, the *word length* is 2 bits. Common word lengths are 4, 8, and 16 bits. Because 8-bit words are frequently used, a special name has been given to this length: 8 bits are called a *byte*. Figure 9-1 shows the dimensions for a small memory. Note that memory length is always a power of 2.

Other important parameters include *access time*, that is, the amount of time needed to fetch a word from memory. For RAMs, the *write* time is also specified. Power supply voltage and current are of concern when heat is a problem or if the equipment is battery-powered.

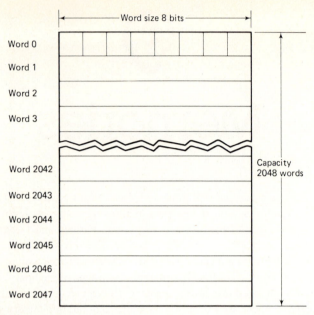

Fig. 9-1 Memory capacity and word size. The capacity is always some power of 2, and here $2^{11} = 2048$.

ROM MEMORY

Read-only memories can be programmed only one time. Because the process of fabricating the ROMs is expensive, only high-volume users find it practical to use these devices. For such applications, costs are low and the ICs can be quite dense. The smallest-sized ROM available has at least 8192 bits. The ROMs are manufactured using Schottky TTL or MOS technology. Table 9-1 lists some examples of typical ROMs. The size of the 6280, for example, is 1024 words of 8 bits (usually written 1024×8). The bipolar ROMs are TTL technology.

Table 9-1 Typical ROMs

Manufacturer and Model	Type	Size	Word Length (Bits)
Monolithic Memories 6280	Bipolar	1024	8
Advanced Micro Devices AM 27S80	Bipolar	1024	8
National Semiconductor DM85S29	Bipolar	1024	8
RCA CDP1833	CMOS	1024	8
Harris HM6312	CMOS	1024	12
Intersil 1M6364	CMOS	4096	8

Because they are so inflexible, ROMs are used less often than are alternate memories. The PROMs include electrically programmable read-only memories (EPROMs), electrically alterable read-only memories (EAROMs), fusible link programmable read-only memories (PROMs), and electrically erasable programmable read-only memories (EEROMs). Each of the types of memories will be described below with examples.

The EPROMs are becoming extremely popular. They can be repeatedly programmed in the field. The memory cells rely on floating-gate MOS transistors to store the information. A floating gate is one that is not connected to an external terminal; it is isolated (floating) in the insulating layer of silicon dioxide. Programming is accomplished by trapping an electric charge on the gate by using a *PROM programmer*. To erase the memory in preparation for changing the contents, it is exposed to ultraviolet (UV) lights for several minutes. A transparent *window* in the chip allows the light to remove the charge on the gate. Two ICs have become standards for EPROMs. The 8K version is the 2708, and the 16K memory is the 2716. Table 9-2 lists some of these memories.

Table 9-2   Typical EPROMs

Manufacturer and Model	Size	Access Time (ns)	Power Supply Voltage (V)
Intel 2708	8K	350,450	12, ±5
TI TMS 27L08	8K	450	12, ±5
Mostek MK2716T	16K	350,390	5
TI TMS 2716	16K	350,450	5
Intel 2716	16K	350,450	5

The EAROMs are based on metal nitride oxide (MNOS) semiconductors and are noted for their low power and good noise tolerance. For this reason, they are frequently found in satellite systems. The EAROMs are slow. It takes 1 to 5 μs to read one word, but they can be reprogrammed in the circuit. The bits can be selectively set to one or zero; therefore, they need not be removed like the EPROMs to change the contents. EAROMs are not manufactured in large quantities, however, because the MNOS production process does not produce high yields. Table 9-3 indicates features of several EAROMs.

Table 9-3   Typical EAROMs

Manufacturer and Model	Size	Word Length (Bits)	Access Time (μs)
Nitron NC 7033	64	4	2–5
GI ER2055	64	8	2–5
Rockwell 10443	256	8	2–6
GI ER2805	2K	4	2–6
Nitron 7810	2K	4	1–4

Fusible link PROMs are programmed by breaking connections between transistors with an external PROM programmer. These links can be made of nichrome, polysilicon, or titanium tungsten. Nichrome fused PROMs are very reliable and require a high-voltage programmer. Both of the other types use low programming voltage, but polysilicon links have sometimes not provided reliable operation. Once information has been stored in these PROMs, it cannot be changed. The most popular technology for these memories is Schottky TTL. The access time to read a word ranges from 45 to 175 ns maximum.

Similar to EPROMs, the EEROMs also depend on a floating gate to store an electrical charge. The major difference is that EEROMs are erased by an electric field instead of UV light. Operational characteristics of EEROMs are about the same as EPROMs in terms of access time and they also must be removed from the circuit to be reprogrammed. Table 9-4 lists some of the EEROMs.

Table 9-4 Examples of EEROMs

Manufacturer and Model	Size	Word Length (Bits)	Power Supply Voltage (V)
Hitachi 48016	2K	8	5
RCA 1843	1K	8	5
NEC μpD454	256	8	12.5
Xicor 2202	1K	1	5

RAMs

Because data in the read/write memories can be changed, power must always be supplied or the data will be erased. Such memories are *volatile* as opposed to *nonvolatile* memories (such as bubble memories) which retain their contents when the equipment is shut off. Families of RAMs are organized around their fabrication technology such as TTL, MOS, and emitter-coupled logic (ECL). The MOS memories can be further subdivided into NMOS and CMOS.

The bipolar TTL and ECL memories are the fastest and also consume the most power. Table 9-5 lists characteristics of some bipolar RAMs. As the table indicates, the ECL devices are the faster of the two.

There are two varieties of CMOS RAMs. *Static* RAMs require a constant power supply voltage, while *dynamic* RAMs permit the power supply to be switched off except for a brief *refresh* period. During the refresh time, memory words are rewritten to maintain the original values. As you would expect, dynamic RAMs require less power. Figure 9-2 shows the difference in the power consumption per bit in the two kinds of memories. Large memories (in excess of 32K) are excellent candidates for dynamic MOS.

Table 9-5 Bipolar RAMs

Logic	Manufacturer and Model	Capacity	Word Size (Bits)	Access Time (ns)	Power (mW)
TTL	TI SN74LS189	16	4	35	525
	Signetics 82S09	64	9	45	1000
	Fairchild 93410	256	1	45	675
	AMD 27S207	256	4	40	600
ECL	Motorola MCM10148	64	1	15	546
	Fujitsu MBM7047	128	1	12	520
	Fairchild F10415	1K	1	20	780
	Siemens GXB100475	1K	4	25	819

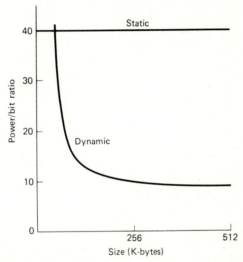

Fig. 9-2 Dynamic memories permit much smaller power supplies to be used.

EXPERIMENT 9-1   INVESTIGATING READ/WRITE MEMORIES

Purpose

To gain experience with RAMs.

Parts

Item	Quantity
7489 RAM	1
LEDs	4
2-kΩ resistor	4
330-Ω resistor	4
SPST switch	9

Procedure

Step 1. Wire the circuit as shown in Fig. 9-3. Note that the memory enable input (pin 2) is grounded; therefore, the RAM is always enabled.

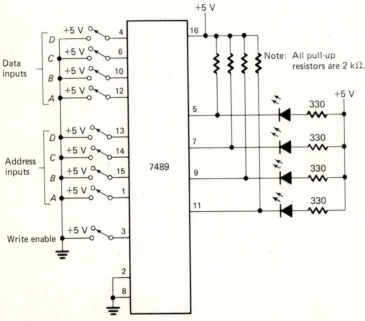

Fig. 9-3 The 7489 read/write memory is a 16-address array of 4-bit words. A total of 64 bits can be stored in the memory.

Step 2. Set the write enable input (pin 3) high, so that the data and address lines can be adjusted without writing.

Step 3. Set all the address bits to ground. This action addresses memory cell zero. Write data into that cell by setting all the data input switches to ground.

Step 4. Write the data into cell zero by grounding the write enable. What is the condition of the LEDs? (*All off*)

160

Step 5. Switch write enable back to 5 V.

Step 6. Continue to write the data in each addressed word as Table 9-1 shows. Record the LED states as each word is written. (Remember that the procedure is to set up the data and address inputs and then ground write enable. Switch write enable back to 5 V before changing addresses or data.)

Step 7. Looking at Table 9-6, how would you interpret data inputs to be related to outputs? *(Outputs are the complement of inputs)*

Table 9-6   Writing into Memory

Address				Data				LED Values				Expected LED Values			
D	*C*	*B*	*A*	*D*	*C*	*B*	*A*	*D*	*C*	*B*	*A*	*D*	*C*	*B*	*A*
L	L	L	H	L	L	L	H					H	H	H	L
L	L	H	L	L	L	H	L					H	H	L	H
L	L	H	H	L	L	H	H					H	H	L	L
L	H	L	L	L	H	L	L					H	L	H	H
L	H	L	H	L	H	L	H					H	L	H	L
L	H	H	L	L	H	H	L					H	L	L	H
L	H	H	H	L	H	H	H					H	L	L	L
H	L	L	L	H	L	L	L					L	H	H	H
H	L	L	H	H	L	L	H					L	H	H	L
H	L	H	L	H	L	H	L					L	H	L	H
H	L	H	H	H	L	H	H					L	H	L	L
H	H	L	L	H	H	L	L					L	L	H	H
H	H	L	H	H	H	L	H					L	L	H	L
H	H	H	L	H	H	H	L					L	L	L	H
H	H	H	H	H	H	H	H					L	L	L	L

Step 8. Now check the contents of the memory by reading each cell in turn. Leave write enable switched to 5 V. Set the address lines for each cell in turn. How do the readouts compare with the outputs in Table 9-6? *(They are the same)*

Ins and Outs

The 7489 memory inputs are 4-bit values that select any address in the range of 0000_2 to 1111_2. Decoding of the address is provided on the chip. The memory output lines are open collector which may be wire ANDed to expand the total memory by combining several 7489 ICs.

The functions of the memory are controlled by the memory enable and write enable as shown in Table 9-7. In electronic equipment with memory, there will be times when the memory should not respond to inputs. This event is especially likely in equipment using a bused architecture. In such cases, the memory is disabled with a high on the enable input.

Table 9-7   7489 Functions

Write Enable	Memory Enable	Action	Output
L	L	Write	Complement of inputs
L	H	Inhibit	Complement of inputs
H	L	Read	Complement of memory data
H	H	Not selected	All high

In writing, the data is set on the lines and the memory address selected. If both memory and write enable are low, the data is stored into that cell. Reading the data is accomplished by enabling the memory and setting the address lines correctly. When the memory is read, the data in the cell is not erased. This type of reading is called *nondestructive*, because memory contents are not altered.

Conclusions

The 7489 is a fully decoded memory with a capacity of 16 words of 4 bits each. Outputs are open collector, so pull-up resistors should be used. The chip is selected by the memory enable input. Writing occurs when write enable is low. Outputs are the complement of the inputs.

EXPERIMENT 9-2   WORKING WITH REGISTER FILES

Purpose

To learn how to use a register file.

Parts

Item	Quantity
74170 register file	1
LEDs	4
2-kΩ resistor	4
330-Ω resistor	4
SPST switch	10

Procedure

Step 1. Connect the circuit as shown in Fig. 9-4.

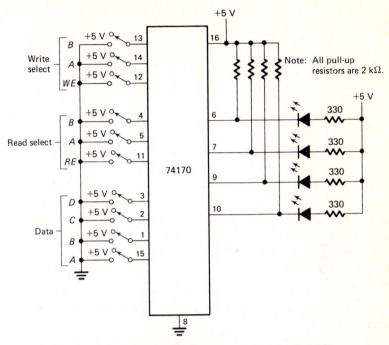

Fig. 9-4 **The 74170 is a 4-by-4 register file. Because read and write addressing are separated, both reading and writing can take place at the same time.**

Step 2. Set both write switches (A and B) low. This selection addresses the first register which has address 00_2. Set the write enable (pin 12) high.

Step 3. Set the data switches to any value you wish to store in the register.

Step 4. Ground the write enable.

Step 5. Check that the value was correctly stored by reading cell 00_2. (Ground both read select lines and the read enable pin.)

Step 6. Store values in cells 01_2 through 11_2 in the same way, and record your results in Table 9-8.

Table 9-8 Writing in the Register File

Write Select		Value Stored				
B	A	D	C	B	A	Value Read
H	L					
L	H					
L	L					

163

Ins and Outs

The register file is very similar to a small memory. It can store four words consisting of 4 bits each. The addresses are decoded on the chip. Separate read and write controls provide the means for simultaneously storing and retrieving the information.

The write functions are shown in Table 9-9. The write enable must be low for any storage operation. The addressing is performed by the write select lines.

Table 9-9 74170 Write Functions

Write Enable	Write Select B	Write Select A	Word* 3	Word* 2	Word* 1	Word* 0
H	X	X	NC	NC	NC	NC
L	L	L	NC	NC	NC	D
L	L	H	NC	NC	D	NC
L	H	L	NC	D	NC	NC
L	H	H	D	NC	NC	NC

*D = data values; NC = no change.

To read the contents in the file, the settings given in Table 9-10 are used. The read enable must be low for normal operation. Each word is addressed just as in the writing process.

Table 9-10 74170 Read Functions

Read Enable	Read Select B	Read Select A	Output
H	X	X	H
L	L	L	Word 0
L	L	H	Word 1
L	H	L	Word 2
L	H	H	Word 3

As the address is placed on the read selection lines, the outputs display the contents of the designated word.

Conclusions

The 74170 register file can store and retrieve four words. The writing and reading operations are individually controlled in contrast to the memory of Experiment 9-1. The separate controls mean that writing does not exclude reading at the same time.

EXPERIMENT 9-3   ADDRESSING A MEMORY

Purpose

To develop an address selection circuit.

Parts

Item	Quantity
7430 NAND	1
7405 hex inverter	1
LED	1
330-Ω resistor	1
SPST switch	1

Procedure

Step 1.   Wire the circuit shown in Fig. 9-5a.

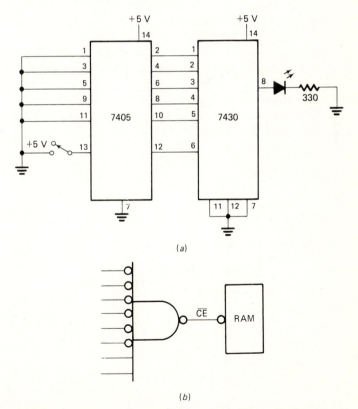

(a)

(b)

Fig. 9-5   The 7430 can be used to decode eight address lines. The inverters decide whether a high or low will activate the selection. *(a)* Schematic; *(b)* logic diagram.

Step 2. With pin 13 of the 7405 connected to 5 V, what is the state of the output? *(High)*

Step 3. Now ground pin 13 of the 7405.

Step 4. How does the output change? *(Low)*

Step 5. Rearrange the inputs to NAND gate by grounding some others and repeat the experiment. Do you find that the output of the circuit is only low for one combination of input levels, no matter how the circuit is arranged?

Ins and Outs

This circuit is an address decoder that would be connected to the memory enable, more often called *chip enable* ($\overline{CE}$), as shown in Fig. 9-5b. The bar over the designation for that input means that a *low* input is the enabling condition.

An address decoder is used to choose just one memory IC out of many. One way this can be done is to use a group of NAND gates as in this experiment, with inverters used in various combinations.

Figure 9-6 shows a small memory decoder network. Only two address lines are used in this example. For each of the four possible address values that can be set, one of the NAND gates will have a low output. If the address lines are $A = 0$ and $B = 1$, NAND gates 0, 1, and 3 have high outputs. These memories will be disabled because the $\overline{CE}$ signal is high. NAND gate 2 will produce a low signal which will enable memory number 2.

The fan-in for the NAND gates is determined by the number of memory chips that must be decoded. In our example each gate had two inputs, and a maximum of four memories could be controlled. In general, the number is

$$N = 2^F$$

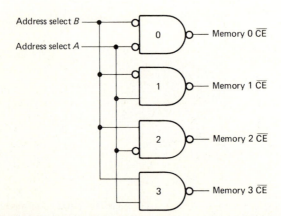

Fig. 9-6   A simplified address decoding network. For any value on the address selection lines, only one NAND gate will produce a low output.

166

where N is the number of memories to decode and F is the fan-in of the NAND gate. In our example the fan-in was 2, so

$$N = 2^2 = 4$$

For the gate used in the experiment, the number of memories is much larger. The fan-in was 8, giving a value of

$$N = 2^8 = 256$$

Conclusions

Memory decoders often use NAND gates to select which RAM will be enabled. By combining inverters with the NAND gates, any memory address can be selected. The fan-in of the NAND gate is the factor which determines how many memories can be decoded.

MAGNETIC BUBBLE MEMORIES

The *bubbles* in these memories are small magnetic domains which are formed into cylinders in thick films of ferrites. Alternatively, the bubbles can also be created in amorphous metal films. In either case, forming the bubbles requires an external magnetic field applied at right angles to the film. Permanent magnets are used to produce this field. Figure 9-7 shows an exploded view of a bubble memory.

The bubbles are moved about in the film by magnetic fields. These fields result from current flowing in the orthogonal coils. The presence of a bubble represents a logic one, and the absence of a bubble represents a zero. The bubbles move in paths

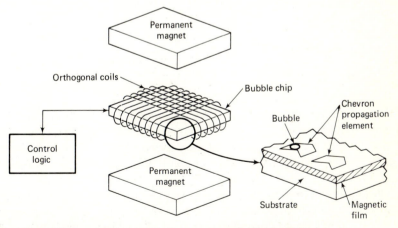

Fig. 9-7 **The bubble memory requires two magnetic fields. Permanent magnets create a field at right angles to the memory plane, while field coils produce magnetic gradients needed to move the bubbles along the propagation path.**

167

formed by Permalloy metal patterns deposited on the film. Typical propagation patterns are shaped like the letters T and Y, chevrons, and asymmetrical chevrons. The patterns act as electromagnets from the induced fields which can attract or repel the bubbles along the path in response to the rotating magnetic field of the coils. As the bubbles move along the Permalloy path, they change the magnetic resistance. By sensing the change, the presence of a bubble can be detected.

All these components are packaged into a single bubble memory device. The Texas Instrument TIB0303, for example, is a 20-pin DIP measuring $1.2 \times 1.2 \times 0.4$ in [3.1 $\times 3.1 \times 1.0$ cm]. Bubble memories are nonvolatile, so turning off the power does not erase the contents. The permanent magnets supply the field necessary to maintain the memory contents.

Magnetic bubbles are organized into loops as shown in Fig. 9-8. The smallest memories need only a simple loop, but practical devices use major and minor loop configurations. The TIB0303, for example, stores 254K bits in 224 minor loops, each holding 1137 bubbles. Because defects may make some of the minor loops unusable, the memories are provided with 28 spare minor loops which can replace the defective ones. The memory controller can tell which loops are good by examining the dedicated minor loop that holds the *memory map*.

As Fig. 9-8 shows, the bubbles move in series along the loops. The write head creates bubbles for the "ones" in the input stream. The read head detects each bubble site as it passes underneath the output station. The replicate head rewrites the bubbles if they are to remain in storage. Table 9-11 compares the characteristics of bubble memories made by Texas Instruments, Rockwell, and Intel.

Table 9-11   Bubble Memories

Manufacturer and Model	Capacity (Bits)	Access Time (ms)	Power (W)	Package Size (DIP Pins)
Texas Instruments TIB0203	92K	4	0.7	14
Texas Instruments TIB0303	254K	7.3	0.9	20
Rockwell RBM256	256K	6	0.8	18
Intel 7110	1024K	40	2.5	20

A complete bubble memory requires the components shown in Fig. 9-9. The controller interfaces with the external equipment (possibly a microprocessor) using the memory. All read, write, and addressing operations are handled by the controller. The controller sends signals to the function timing generator which accesses the memory.

The function driver converts digital control signals to current pulses for generating, replicating, or transferring bubbles. The coil drivers and diode arrays generate two triangular current drives, 90° out of phase with each other for the orthogonal coils. The small-output signal of the memory (about 3 mV) is converted to TTL levels by the *RC* network and the sense amplifier. Table 9-12 lists some of the features of the auxiliary chips used with the Texas Instrument bubble memory.

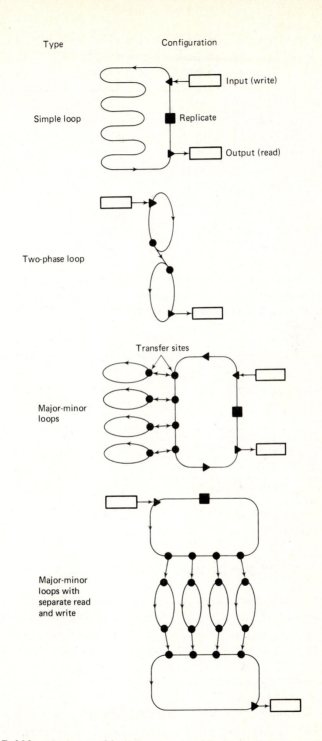

Fig. 9-8 **Bubble memory architectures are combinations of major and minor loops.**

169

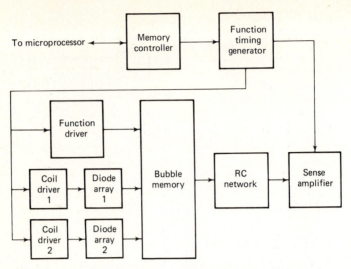

Fig. 9-9 Timing and access of a bubble memory depend on the memory controller and function timing generator. Other external chips drive the coils and sense the output.

Table 9-12   Texas Instrument Bubble Memory Auxiliary Devices

Device	Power Supply Voltage (V)	Package Size (DIP Pins)	Features
74LS361 function timing generator	5	22	12-megahertz (MHz) clock frequency
75281 sense amplifier	±5	14	TTL-compatible inputs, three-state outputs
75380 function driver	+5, −12	16	TTL-compatible inputs
75382 coil driver	+5, −12	14	TTL-compatible inputs, totem-pole outputs

CHARGE-COUPLED DEVICE MEMORIES

Charge-coupled device (CCD) memories are also suited for large amounts of storage in a small area. Unlike bubble memories, CCDs are volatile, but CCDs do not require the bulky permanent magnets and thus can be packaged into a smaller volume. Two basic structures are used to organize CCD memories. Synchronous structures move the bits in lock-step, while nonsynchronous memories operate as independent serial loops. Figure 9-10 shows the synchronous serpentine and nonsynchronous serial-parallel-serial organizations.

Data is stored in the CCD memory as charged (ones) or uncharged (zero) potential wells under gates in a long register. The charge packets move along the register to circulate the data pattern. At the end of the register, data can be read out or *recirculated* to store the charges back at the beginning of the register.

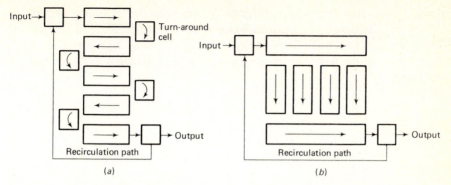

Fig. 9-10   The serpentine CCD memory *(a)* is a synchronous organization, while the serial-parallel-serial memory *(b)* is nonsynchronous.

MEMORY COMPARISON

You may be wondering why so many different kinds of memory are necessary. Figure 9-11 shows how each type of device occupies a unique area of the speed/capacity chart. Most competitive are bubble and CCD memories. They are closely related in

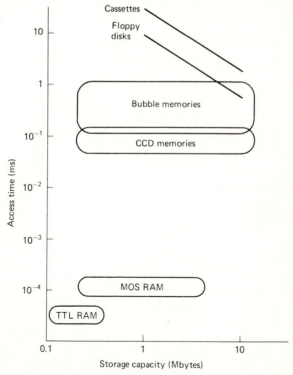

Fig. 9-11   The storage capacity and access time for a variety of memories are shown in this chart. Normally high speed implies less storage capacity, and conversely.

capacity and speed, and seem to be growing even more similar as new products are announced. Because they are priced so closely alike, many people think that one will overwhelm the other and occupy that spot on the graph alone. As a preview of things to come, note the large gap on the chart between RAMs and the other types of memories.

10

Solid-State Lamps and Displays

The experiments in this chapter will investigate the LED lamps that you used in earlier experiments. Consideration will also be given to seven-segment LED displays. These devices allow us to show numeric values rather than simple binary indications. The decoder/drivers used with these displays simplify the task of converting binary to decimal digits. In addition, the decoder/drivers supply sufficient current to produce a bright, easily read value on the displays.

EXPERIMENT 10-1   EXAMINING LED LAMPS

Purpose

To review the operation of discrete LEDs.

Parts

Item	Quantity
LED	1
330-Ω resistor	1
1-kΩ resistor	1
Voltmeter	1

Procedure

Step 1.   Wire the circuit shown in Fig. 10-1a. Is the LED on or off? *(On)* Measure the voltage drop across the resistor. *(About 3 V)*

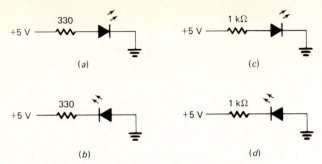

Fig. 10-1 **LED circuits demonstrate the diode characteristics and the current drawn by these devices.**

Step 2. Compute the current in the circuit

$$I = \frac{V}{R} = \frac{V}{330}$$

where V is the resistor voltage drop measured in Step 1. *(About 9 mA)*

Step 3. Reverse the LED as shown in Fig. 10-1*b*. Is the LED off or on? *(Off)*

Step 4. Change the circuit to the one shown in Fig. 10-1*c*. Is the LED brighter or dimmer than in the case of the circuit in Fig. 10-1*a*. *(Dimmer)* Again measure the resistor voltage drop. *(About 3 V)*

Step 5. Reverse the LED as in Fig. 10-1*d*. Is it on or off? *(Off)*

Ins and Outs

The LED works like a regular diode in that it must be forward-biased to illuminate. That condition is satisfied for the circuits shown in Figs. 10-1*a* and 10-1*c*. When the LED is on, it will drop 1.8 V [typical for a red gallium arsenide phosphide LED]. Current through the device should be limited to the range of 5 to 40 mA for minimum power drain and also for adequate light intensity.

Conclusions

The LED must be forward-biased to illuminate. A current-limiting resistor in the range of 100 to 500 Ω should be connected in series with the LED.

EXPERIMENT 10-2   CONTROLLING A SEVEN-SEGMENT COMMON-ANODE DISPLAY

Purpose

To become familiar with seven-segment LED displays.

Parts

Item	Quantity
IEE 1712 display	1
330-Ω resistor	6

Procedure

Step 1. Construct the circuit shown in Fig. 10-2. What digit appears on the display? *(Zero)*

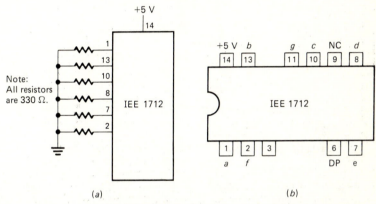

(a)   *(b)*

Fig. 10-2 The common-anode display uses a single power supply pin and grounds the pins of segments to be illuminated. (Either pin 14 or 3 can be connected to the power supply of the IEE 1712.) *(a)* Circuit; *(b)* pin assignment.

Steps 2 through 10. Continue to test the IEE 1712 for each case listed in Table 10-1. Record the value displayed and compare it to the expected digit.

Table 10-1   IEE 1712 Display

Step	Pins Grounded through Resistors	Display	Expected Value
2	13, 10		1
3	1, 13, 8, 7, 11		2
4	1, 13, 10, 8, 11		3
5	13, 10, 2, 11		4
6	1, 10, 8, 2, 11		5
7	10, 8, 7, 2, 11		6
8	1, 13, 10		7
9	1, 13, 10, 8, 7, 2, 11		8
10	1, 13, 10, 2, 11		9

Ins and Outs

The common-anode display consists of seven LEDs arranged in the pattern shown in Fig. 10-3. Each segment, or bar, used to display a portion of the digit is a separate LED. All these diodes have their anodes tied to a common point (pin 14) which is connected to the power supply. With this configuration, grounding the cathode of any LED will cause that segment to glow. Table 10-2 lists the segments used to represent each decimal digit. (Note that the IEE 1712 also has a decimal point LED in addition to the seven segments for the digits.)

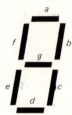

Fig. 10-3 The seven LEDs comprising the display are identified by letters.

Table 10-2 Digit to Segment Cross-Reference List

Digit	Display	a	b	c	d	e	f	g
0		X	X	X	X	X	X	
1			X	X				
2		X	X		X	X		X
3		X	X	X	X			X
4			X	X			X	X
5		X		X	X		X	X
6				X	X	X	X	X
7		X	X	X				
8		X	X	X	X	X	X	X
9		X	X	X			X	X

Conclusions

The common-anode LED has a single power supply pin tied to the anode of every diode. Grounding each cathode lights the corresponding segment. A cross-reference table, such as Table 10-2, indicates which segments should be on for each digit.

EXPERIMENT 10-3 COMPARING A SEVEN-SEGMENT COMMON-CATHODE DISPLAY

Purpose

To examine a common-cathode LED display.

Parts

Item	Quantity
MAN 74 common-cathode display	1
330-Ω resistor	6

Procedure

Step 1. Connect the circuit shown in Fig. 10-4.

Step 2. What digit is displayed? *(Zero)*

Steps 3 through 11. Change the pull-up resistor connections as listed in Table 10-3. Compare your results with those expected.

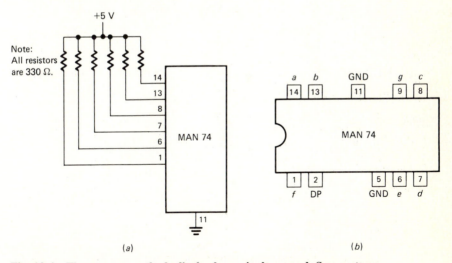

Fig. 10-4 The common-cathode display has a single ground. Segments are illuminated by tying them to the power supply through a pull-up resistor. (Either pin 5 or 11 can be used as the ground). *(a)* Circuit; *(b)* pin assignments.

Table 10-3   MAN 74 Display

Step	Pins Connected to Pull-up Resistors	Display	Expected Value
3	13, 8		1
4	14, 13, 7, 6, 8		2
5	14, 13, 8, 7, 9		3
6	13, 8, 1, 9		4
7	14, 8, 7, 1, 9		5
8	8, 7, 6, 1, 9		6
9	14, 13, 8		7
10	14, 13, 8, 7, 6, 1, 9		8
11	14, 13, 8, 1, 9		9

Ins and Outs

The internal connections of the MAN 74 are just the reverse of the IEE 1712. All the cathodes are joined to the common ground, pin 11. Each segment is selected by attaching the appropriate pin to the 5-V supply through a resistor. The segments for each digit are the same as those listed in Table 10-2.

Conclusions

Seven-segment displays are available in either common-anode or common-cathode configurations. The choice of which display is more convenient to use depends on the decoder/driver, which will be investigated in the following experiments.

EXPERIMENT 10-4   DRIVING A COMMON-ANODE DISPLAY

Purpose

To build a BCD to seven-segment display decoding circuit.

Parts

Item	Quantity
7447 common-anode decoder/driver	1
IEE 1712 display	1
330-Ω resistors	7
SPDT switch	4

Procedure

Step 1.   Wire the circuit shown in Fig. 10-5.

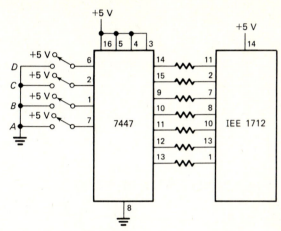

Note: All resistors are 330 Ω.

Fig. 10-5   The 7447 converts BCD inputs to signals which ground the cathodes of the segments to display each digit.

Step 2.   Ground all the switches. What does the display show? *(Zero)*

Step 3.   Attach switch *A* to 5 V. What digit appears on the display? *(One)*

Steps 4 through 11.   Continue to use the switches to generate the remaining digits as listed in Table 10-4.

Table 10-4   Common-Anode Decoder/Driver Experiment

Step	Switch				Display
	D	*C*	*B*	*A*	
4	0 V	0 V	5 V	0 V	2
5	0	0	5	5	3
6	0	5	0	0	4
7	0	5	0	5	5
8	0	5	5	0	6
9	0	5	5	5	7
10	5	0	0	0	8
11	5	0	0	5	9

179

Ins and Outs

The 7447 is a combination of a decoding circuit with the logic needed to create grounds for the various segments in each digit. In effect, the 7447 implements the cross-reference list in Table 10-2.

Conclusions

BCD inputs can be converted into signals which can drive a common-anode display.

EXPERIMENT 10-5 DRIVING A COMMON-CATHODE DISPLAY

Purpose

To construct a circuit which can drive a common-cathode display using a BCD input.

Parts

Item	Quantity
7448 common-cathode decoder/driver	1
MAN 74 display	1
330-Ω resistor	7
SPDT switch	4

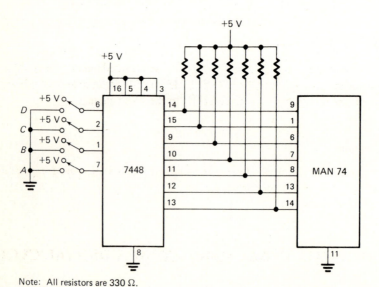

Note: All resistors are 330 Ω.

Fig. 10-6 The 7448 decoder/driver is used with common-cathode displays to change BCD inputs into the digits on the display.

180

Procedure

Step 1.   Construct the circuit shown in Fig. 10-6.

Step 2.   Ground all the switches. Verify that a zero appears on the display.

Step 3.   What effect do you predict converting switch *A* to 5 V will have? *(Display a one)* Check your prediction.

Steps 4 through 11.   Change the switch settings as listed in Table 10-5.

Table 10-5   **Common-Cathode Decoder/Driver Experiment**

Step	D	C	B	A	Display
		Switch			
4	0 V	0 V	5 V	0 V	2
5	0	0	5	5	3
6	0	5	0	0	4
7	0	5	0	5	5
8	0	5	5	0	6
9	0	5	5	5	7
10	5	0	0	0	8
11	5	0	0	5	9

Ins and Outs

The 7448 decoder/driver works in a manner practically identical to that of the 7447. The major difference is that the 7448 provides the 5-V input to appropriate segments to illuminate the digit which corresponds to the BCD input.

Conclusions

If a common-cathode LED display is being used in a BCD coded application, the 7448 is the proper decoder/driver for it.

EXPERIMENT 10-6   SIMULATING A DIGITAL CLOCK

Purpose

To combine timing with a display circuit.

Parts

Item	Quantity
7447 decoder/driver	1
IEE 1712 display	1
7490 counter	1
555 timer	1
330-Ω resistor	7
15-kΩ resistor	1
68-kΩ resistor	1
10-μF capacitor	1

Procedure

Step 1. You may first wish to review the experiments covering the 555 timer and the 7490 counter.

Step 2. Wire the 555 as shown in Fig. 10-7. Check the output with a logic probe. The level should shift between high and low at about a 1-Hz rate.

Step 3. Add the 7490 to the circuit. With a logic probe, verify that the output from pins 1 and 12 alternates with half the frequency of the timer.

Step 4. Complete the circuit. The display should increment from zero through nine and then repeat.

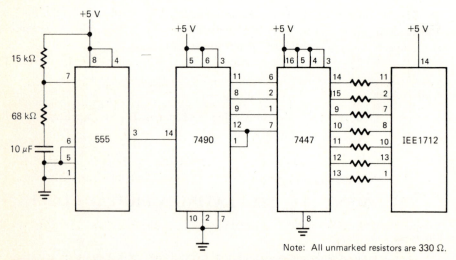

Note: All unmarked resistors are 330 Ω.

Fig. 10-7 Combinations of a timer, counter, and display system demonstrate the key elements of a digital clock.

182

Ins and Outs

The 555 with the components selected for this experiment generates a 1-Hz square wave. The pulses are totaled by the counter which provides a BCD output. The BCD numbers are, in turn, decoded by the 7447, which also drives the display.

A digital clock differs from this circuit mostly in the accuracy of the timer used. Either a crystal oscillator or the 60-Hz line frequency provide a more accurate time base in a clock than is possible with use of the 555. The remainder of this circuit, however, provides quite precise timing and display of the incoming pulses.

Conclusions

The circuits examined individually in earlier experiments can be connected together as a simple clock. The accuracy of this clock is not very good because the resistors and the capacitor connected to the 555 have large tolerances.

EXPERIMENT 10-7   IMPROVING THE LOGIC PROBE

Purpose

To build a better version of the logic probe.

Parts

Item	Quantity
7404 hex inverter	1
LEDs	2
180-Ω resistor	2

Procedure

Step 1.   Wire the circuit shown in Fig. 10-8.

Step 2.   Touch the probe to ground. Which LED indicator is on? *(The low LED)*

Step 3.   Touch the probe to the supply voltage. Does the high LED illuminate? *(Yes)*

Ins and Outs

The low LED is connected through two inverters in series to the probe tip. The LED will light whenever the probe tip is low, because the two inverters produce a low level on the cathode of the LED.

The high LED is connected to the probe tip through a single inverter; thus the cathode of that LED will be at a low level when the probe touches a high signal. In such cases, the LED is illuminated.

Other advantages of this probe include the reduced loading that it introduces to the circuit under test. The load is only two TTL inverters, which are less than the series

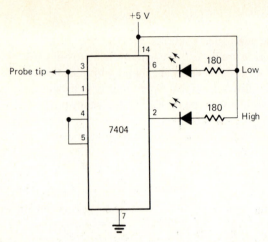

Fig. 10-8 **The improved logic probe reduces problems associated with circuit loading. The new version eliminates that problem and also indicates in both the high and low states.**

resistor and the LED. This probe also presents an alternate blinking pattern for low-frequency pulse trains.

Conclusions

By adding inverters to the basic logic probe, an improved design results. The new probe indicates high or low levels, reduces loading, and blinks for pulse trains.

APPENDIX A
54/74 Families of Compatible TTL Circuits: Data Sheets

You will frequently find a code appended to the number of the integrated circuit (such as SN7405, CD7405, or LM7405). All of these prefixes merely identify the manufacturer but do not change the function performed. Chips with various prefixes can be interchanged freely in these experiments.

Integrated circuit pin assignments and functional descriptions (Courtesy of Texas Instruments, Inc.)

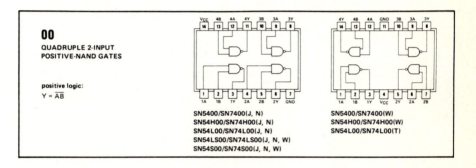

00

QUADRUPLE 2-INPUT
POSITIVE-NAND GATES

positive logic:
$Y = \overline{AB}$

SN5400/SN7400(J, N)
SN54H00/SN74H00(J, N)
SN54L00/SN74L00(J, N)
SN54LS00/SN74LS00(J, N, W)
SN54S00/SN74S00(J, N, W)

SN5400/SN7400(W)
SN54H00/SN74H00(W)
SN54L00/SN74L00(T)

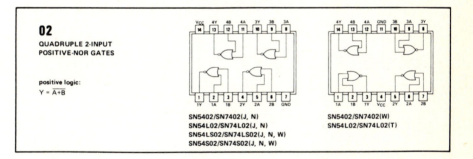

02

QUADRUPLE 2-INPUT
POSITIVE-NOR GATES

positive logic:
$Y = \overline{A+B}$

SN5402/SN7402(J, N)
SN54L02/SN74L02(J, N)
SN54LS02/SN74LS02(J, N, W)
SN54S02/SN74S02(J, N, W)

SN5402/SN7402(W)
SN54L02/SN74L02(T)

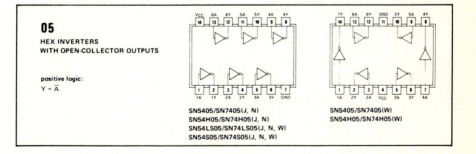

05

HEX INVERTERS
WITH OPEN-COLLECTOR OUTPUTS

positive logic:
$Y = \overline{A}$

SN5405/SN7405(J, N)
SN54H05/SN74H05(J, N)
SN54LS05/SN74LS05(J, N, W)
SN54S05/SN74S05(J, N, W)

SN5405/SN7405(W)
SN54H05/SN74H05(W)

QUADRUPLE 2-INPUT
POSITIVE-AND GATES

08

positive logic:
Y = AB

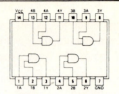

SN5408 (J, W)   SN7408 (J, N)
SN54LS08 (J, W)   SN74LS08 (J, N)
SN54S08 (J, W)   SN74S08 (J, N)

QUADRUPLE 2-INPUT
POSITIVE-AND GATES
WITH OPEN-COLLECTOR OUTPUTS

09

positive logic:
Y = AB

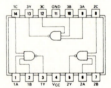

SN5409 (J, W)   SN7409 (J, N)
SN54LS09 (J, W)   SN74LS09 (J, N)
SN54S09 (J, W)   SN74S09 (J, N)

TRIPLE 3-INPUT
POSITIVE-NAND GATES

10

positive logic:
Y = $\overline{ABC}$

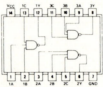

SN5410 (J)   SN7410 (J, N)   SN5410 (W)
SN54H10 (J)   SN74H10 (J, N)   SN54H10 (W)
SN54L10 (J)   SN74L10 (J, N)   SN54L10 (T)
SN54LS10 (J, W)   SN74LS10 (J, N)
SN54S10 (J, W)   SN74S10 (J, N)

11

TRIPLE 3-INPUT
POSITIVE-AND GATES

positive logic:
Y = ABC

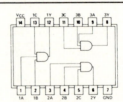

SN54H11/SN74H11(J, N)   SN54H11/SN74H11(W)
SN54LS11/SN74LS11(J, N, W)
SN54S11/SN74S11(J, N, W)

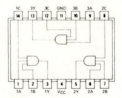

188

14

HEX SCHMITT-TRIGGER INVERTERS

positive logic:
$Y = \overline{A}$

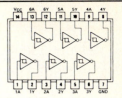

SN5414/SN7414(J, N, W)

23

EXPANDABLE DUAL 4-INPUT POSITIVE-NOR GATES WITH STROBE

positive logic:
$1Y = \overline{1G(1A+1B+1C+1D)+X}$
$2Y = \overline{2G(2A+2B+2C+2D)}$
    X = output of SN5460/SN7460

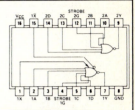

SN5423/SN7423 (J, N, W)

32

QUADRUPLE 2-INPUT POSITIVE-OR GATES

positive logic:
$Y = A+B$

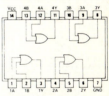

SN5432/SN7432(J, N, W)
SN54LS32/SN74LS32(J, N, W)

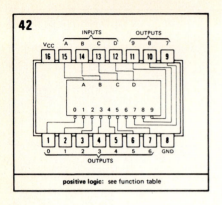

42

positive logic: see function table

FUNCTION TABLE

NO.	'42A, 'L42 BCD INPUT				'43A, 'L43 EXCESS-3-INPUT				'44A, 'L44 EXCESS-3-GRAY INPUT				ALL TYPES DECIMAL OUTPUT									
	D	C	B	A	D	C	B	A	D	C	B	A	0	1	2	3	4	5	6	7	8	9
0	L	L	L	L	L	L	H	H	L	L	H	L	L	H	H	H	H	H	H	H	H	H
1	L	L	L	H	L	H	L	L	L	H	H	L	H	L	H	H	H	H	H	H	H	H
2	L	L	H	L	L	H	L	H	L	H	H	H	H	H	L	H	H	H	H	H	H	H
3	L	L	H	H	L	H	H	L	L	H	L	H	H	H	H	L	H	H	H	H	H	H
4	L	H	L	L	L	H	H	H	L	H	L	L	H	H	H	H	L	H	H	H	H	H
5	L	H	L	H	H	L	L	L	H	H	L	L	H	H	H	H	H	L	H	H	H	H
6	L	H	H	L	H	L	L	H	H	H	L	H	H	H	H	H	H	H	L	H	H	H
7	L	H	H	H	H	L	H	L	H	H	H	H	H	H	H	H	H	H	H	L	H	H
8	H	L	L	L	H	L	H	H	H	H	H	L	H	H	H	H	H	H	H	H	L	H
9	H	L	L	H	H	H	L	L	H	L	H	L	H	H	H	H	H	H	H	H	H	L
INVALID	H	L	H	L	H	H	L	H	H	L	H	H	H	H	H	H	H	H	H	H	H	H
INVALID	H	L	H	H	H	H	H	L	H	L	L	H	H	H	H	H	H	H	H	H	H	H
INVALID	H	H	L	L	H	H	H	H	H	L	L	L	H	H	H	H	H	H	H	H	H	H
INVALID	H	H	L	H	L	L	L	L	L	L	L	L	H	H	H	H	H	H	H	H	H	H
INVALID	H	H	H	L	L	L	L	H	L	L	L	H	H	H	H	H	H	H	H	H	H	H
INVALID	H	H	H	H	L	L	H	L	L	L	H	H	H	H	H	H	H	H	H	H	H	H

H = high level, L = low level

190

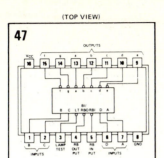

(TOP VIEW)

47

SEGMENT
IDENTIFICATION

NUMERICAL DESIGNATIONS AND RESULTANT DISPLAYS

'46A, '47A, 'L46, 'L47, 'LS47 FUNCTION TABLE

DECIMAL OR FUNCTION	INPUTS						BI/RBO†	OUTPUTS							NOTE
	LT	RBI	D	C	B	A		a	b	c	d	e	f	g	
0	H	H	L	L	L	L	H	ON	ON	ON	ON	ON	ON	OFF	
1	H	X	L	L	L	H	H	OFF	ON	ON	OFF	OFF	OFF	OFF	
2	H	X	L	L	H	L	H	ON	ON	OFF	ON	ON	OFF	ON	
3	H	X	L	L	H	H	H	ON	ON	ON	ON	OFF	OFF	ON	
4	H	X	L	H	L	L	H	OFF	ON	ON	OFF	OFF	ON	ON	
5	H	X	L	H	L	H	H	ON	OFF	ON	ON	OFF	ON	ON	
6	H	X	L	H	H	L	H	OFF	OFF	ON	ON	ON	ON	ON	
7	H	X	L	H	H	H	H	ON	ON	ON	OFF	OFF	OFF	OFF	1
8	H	X	H	L	L	L	H	ON	ON	ON	ON	ON	ON	ON	
9	H	X	H	L	L	H	H	ON	ON	ON	OFF	ON	ON	ON	
10	H	X	H	L	H	L	H	OFF	OFF	OFF	ON	ON	OFF	ON	
11	H	X	H	L	H	H	H	OFF	OFF	ON	ON	OFF	OFF	ON	
12	H	X	H	H	L	L	H	OFF	ON	OFF	OFF	OFF	ON	ON	
13	H	X	H	H	L	H	H	ON	OFF	OFF	ON	OFF	ON	ON	
14	H	X	H	H	H	L	H	OFF	OFF	OFF	ON	ON	ON	ON	
15	H	X	H	H	H	H	H	OFF	OFF	OFF	OFF	OFF	OFF	OFF	
BI	X	X	X	X	X	X	L	OFF	OFF	OFF	OFF	OFF	OFF	OFF	2
RBI	H	L	L	L	L	L	L	OFF	OFF	OFF	OFF	OFF	OFF	OFF	3
LT	L	X	X	X	X	X	H	ON	ON	ON	ON	ON	ON	ON	4

H = high level, L = low level, X = irrelevant

NOTES: 1. The blanking input (BI) must be open or held at a high logic level when output functions 0 through 15 are desired. The ripple-blanking input (RBI) must be open or high if blanking of a decimal zero is not desired.

2. When a low logic level is applied directly to the blanking input (BI), all segment outputs are off regardless of the level of any other input.

3. When ripple-blanking input (RBI) and Inputs A, B, C, and D are at a low level with the lamp test input high, all segment outputs go off and the ripple-blanking output (RBO) goes to a low level (response condition).

4. When the blanking input/ripple blanking output (BI/RBO) is open or held high and a low is applied to the lamp-test input, all segment outputs are on.

†BI/RBO is wire-AND logic serving as blanking input (BI) and/or ripple-blanking output (RBO).

(TOP VIEW)

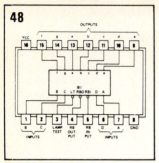

FUNCTION TABLE

DECIMAL OR FUNCTION	INPUTS						BI/RBO†	OUTPUTS							NOTE
	LT	RBI	D	C	B	A		a	b	c	d	e	f	g	
0	H	H	L	L	L	L	H	H	H	H	H	H	H	L	1
1	H	X	L	L	L	H	H	L	H	H	L	L	L	L	1
2	H	X	L	L	H	L	H	H	H	L	H	H	L	H	
3	H	X	L	L	H	H	H	H	H	H	H	L	L	H	
4	H	X	L	H	L	L	H	L	H	H	L	L	H	H	
5	H	X	L	H	L	H	H	H	L	H	H	L	H	H	
6	H	X	L	H	H	L	H	L	L	H	H	H	H	H	
7	H	X	L	H	H	H	H	H	H	H	L	L	L	L	
8	H	X	H	L	L	L	H	H	H	H	H	H	H	H	
9	H	X	H	L	L	H	H	H	H	H	L	L	H	H	
10	H	X	H	L	H	L	H	L	L	L	H	H	L	H	
11	H	X	H	L	H	H	H	L	L	H	H	L	L	H	
12	H	X	H	H	L	L	H	L	H	L	L	L	H	H	
13	H	X	H	H	L	H	H	H	L	L	H	L	H	H	
14	H	X	H	H	H	L	H	L	L	L	H	H	H	H	
15	H	X	H	H	H	H	H	L	L	L	L	L	L	L	
BI	X	X	X	X	X	X	L	L	L	L	L	L	L	L	2
RBI	H	L	L	L	L	L	L	L	L	L	L	L	L	L	3
LT	L	X	X	X	X	X	H	H	H	H	H	H	H	H	4

H = high level, L = low level, X = irrelevant

NOTES: 1. The blanking input (BI) must be open or held at a high logic level when output functions 0 through 15 are desired. The ripple-blanking input (RBI) must be open or high, if blanking of a decimal zero is not desired.

2. When a low logic level is applied directly to the blanking input (BI), all segment outputs are low regardless of the level of any other input.

3. When ripple-blanking input (RBI) and inputs A, B, C, and D are at a low level with the lamp-test input high, all segment outputs go low and the ripple-blanking output (RBO) goes to a low level (response condition).

4. When the blanking input/ripple-blanking output (BI/RBO) is open or held high and a low is applied to the lamp-test input, all segment outputs are high.

†BI/RBO is wire-AND logic serving as blanking input (BI) and/or ripple-blanking output (RBO).

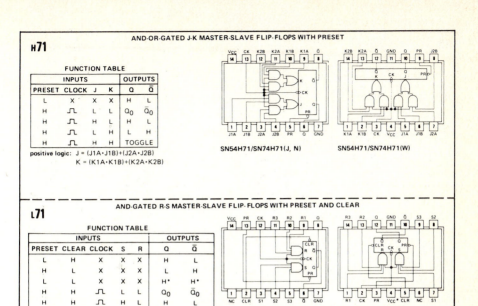

H71 — AND-OR-GATED J-K MASTER-SLAVE FLIP-FLOPS WITH PRESET

FUNCTION TABLE

INPUTS				OUTPUTS	
PRESET	CLOCK	J	K	Q	Q̄
L	X	X	X	H	L
H	⊓	L	L	Q_0	$\bar{Q}_0$
H	⊓	H	L	H	L
H	⊓	L	H	L	H
H	⊓	H	H	TOGGLE	

positive logic: J = (J1A·J1B)+(J2A·J2B)
K = (K1A·K1B)+(K2A·K2B)

SN54H71/SN74H71(J, N)   SN54H71/SN74H71(W)

L71 — AND-GATED R-S MASTER-SLAVE FLIP-FLOPS WITH PRESET AND CLEAR

FUNCTION TABLE

INPUTS					OUTPUTS	
PRESET	CLEAR	CLOCK	S	R	Q	Q̄
L	H	X	X	X	H	L
H	L	X	X	X	L	H
L	L	X	X	X	H*	H*
H	H	⊓	L	L	Q_0	$\bar{Q}_0$
H	H	⊓	H	L	H	L
H	H	⊓	L	H	L	H
H	H	⊓	H	H	INDETERMINATE	

positive logic: R = R1·R2·R3
S = S1·S2·S3

SN54L71/SN74L71(J, N)   SN54L71/SN74L71(T)

NC—No internal connection

* This configuration is nonstable; that is, it will not persist when preset and clear inputs return to their inactive (high) level.

74 — DUAL D-TYPE POSITIVE-EDGE-TRIGGERED FLIP-FLOPS WITH PRESET AND CLEAR

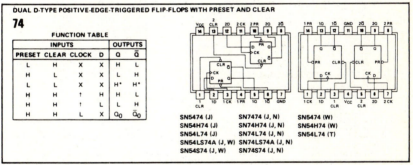

FUNCTION TABLE

INPUTS				OUTPUTS	
PRESET	CLEAR	CLOCK	D	Q	Q̄
L	H	X	X	H	L
H	L	X	X	L	H
L	L	X	X	H*	H*
H	H	↑	H	H	L
H	H	↑	L	L	H
H	H	L	X	Q_0	$\bar{Q}_0$

SN5474 (J)	SN7474 (J, N)	SN5474 (W)
SN54H74 (J)	SN74H74 (J, N)	SN54H74 (W)
SN54L74 (J)	SN74L74 (J, N)	SN54L74 (T)
SN54LS74A (J, W)	SN74LS74A (J, N)	
SN54S74 (J, W)	SN74S74 (J, N)	

* This configuration is nonstable; that is, it will not persist when preset and clear inputs return to their inactive (high) level.

76 — DUAL J-K FLIP-FLOPS WITH PRESET AND CLEAR

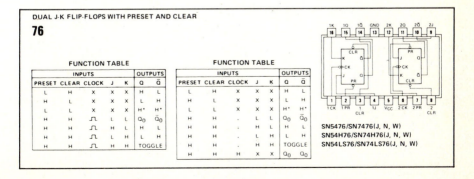

FUNCTION TABLE

INPUTS					OUTPUTS	
PRESET	CLEAR	CLOCK	J	K	Q	Q̄
L	H	X	X	X	H	L
H	L	X	X	X	L	H
L	L	X	X	X	H*	H*
H	H	⊓	L	L	Q_0	$\bar{Q}_0$
H	H	⊓	H	L	H	L
H	H	⊓	L	H	L	H
H	H	⊓	H	H	TOGGLE	

FUNCTION TABLE

INPUTS					OUTPUTS	
PRESET	CLEAR	CLOCK	J	K	Q	Q̄
L	H	X	X	X	H	L
H	L	X	X	X	L	H
L	L	X	X	X	H*	H*
H	H	⌐	L	L	Q_0	$\bar{Q}_0$
H	H	⌐	H	L	H	L
H	H	⌐	L	H	L	H
H	H	⌐	H	H	TOGGLE	
H	H	H	X	X	Q_0	$\bar{Q}_0$

SN5476/SN7476(J, N, W)
SN54H76/SN74H76(J, N, W)
SN54LS76/SN74LS76(J, N, W)

FUNCTION TABLE

INPUTS				OUTPUTS					
				WHEN C0 = L			WHEN C0 = H		
A1	B1	A2	B2	Σ1	Σ2	C2	Σ1	Σ2	C2
L	L	L	L	L	L	L	H	L	L
H	L	L	L	H	L	L	L	H	L
L	H	L	L	H	L	L	L	H	L
H	H	L	L	L	H	L	H	H	L
L	L	H	L	L	H	L	H	H	L
H	L	H	L	H	H	L	L	L	H
L	H	H	L	H	H	L	L	L	H
H	H	H	L	L	L	H	H	L	H
L	L	L	H	L	H	L	H	H	L
H	L	L	H	H	H	L	L	L	H
L	H	L	H	H	H	L	L	L	H
H	H	L	H	L	L	H	H	L	H
L	L	H	H	L	L	H	H	L	H
H	L	H	H	H	L	H	L	H	H
L	H	H	H	H	L	H	L	H	H
H	H	H	H	L	H	H	H	H	H

H = high level, L = low level

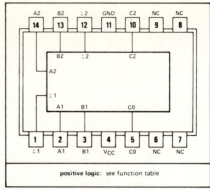

positive logic: see function table

NC–No internal connection

description

These full adders perform the addition of two 2-bit binary numbers. The sum (Σ) outputs are provided for each bit and the resultant carry (C2) is obtained from the second bit. Designed for medium-to-high-speed, multiple-bit, parallel-add/serial-carry applications, these circuits utilize high-speed, high-fan-out transistor-transistor logic (TTL) and are compatible with both DTL and TTL logic families. The implementation of a single-inversion, high-speed, Darlington-connected serial-carry circuit within each bit minimizes the necessity for extensive "look-ahead" and carry-cascading circuits.

functional block diagram

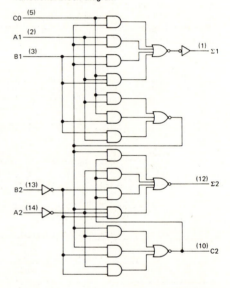

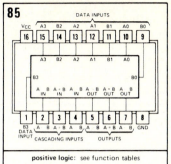

'85, 'S85
J OR N DUAL-IN-LINE OR
W FLAT PACKAGE (TOP VIEW)

positive logic: see function tables

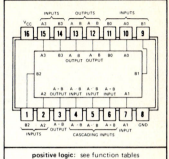

'L85
J OR N
DUAL-IN-LINE PACKAGE (TOP VIEW)

positive logic: see function tables

description

These four-bit magnitude comparators perform comparison of straight binary and straight BCD (8-4-2-1) codes. Three fully decoded decisions about two 4-bit words (A, B) are made and are externally available at three outputs. These devices are fully expandable to any number of bits without external gates. Words of greater length may be compared by connecting comparators in cascade. The A > B, A < B, and A = B outputs of a stage handling less-significant bits are connected to the corresponding A > B, A < B, and A = B inputs of the next stage handling more-significant bits. The stage handling the least-significant bits must have a high-level voltage applied to the A = B input and additionally for the 'L85, low-level voltages applied to the A > B and A < B inputs. The cascading paths of the '85 and 'S85 are implemented with only a two-gate-level delay to reduce overall comparison times for long words.

FUNCTION TABLES

COMPARING INPUTS				CASCADING INPUTS			OUTPUTS		
A3, B3	A2, B2	A1, B1	A0, B0	A > B	A < B	A = B	A > B	A < B	A = B
A3 > B3	X	X	X	X	X	X	H	L	L
A3 < B3	X	X	X	X	X	X	L	H	L
A3 = B3	A2 > B2	X	X	X	X	X	H	L	L
A3 = B3	A2 < B2	X	X	X	X	X	L	H	L
A3 = B2	A2 = B2	A1 > B1	X	X	X	X	H	L	L
A3 = B3	A2 = B2	A1 < B1	X	X	X	X	L	H	L
A3 = B3	A2 = B2	A1 = B1	A0 > B0	X	X	X	H	L	L
A3 = B3	A2 = B2	A1 = B1	A0 < B0	X	X	X	L	H	L
A3 = B3	A2 = B2	A1 = B1	A0 = B0	H	L	L	H	L	L
A3 = B3	A2 = B2	A1 = B1	A0 = B0	L	H	L	L	H	L
A3 = B3	A2 = B2	A1 = B1	A0 = B0	L	L	H	L	L	H

'85, 'S85

A3, B3	A2, B2	A1, B1	A0, B0	A > B	A < B	A = B	A > B	A < B	A = B
A3 = B3	A2 = B2	A1 = B1	A0 = B0	X	X	H	L	L	H
A3 = B3	A2 = B2	A1 = B1	A0 = B0	H	H	L	L	L	L
A3 = B3	A2 = B2	A1 = B1	A0 = B0	L	L	L	H	H	L

'L85

A3, B3	A2, B2	A1, B1	A0, B0	A > B	A < B	A = B	A > B	A < B	A = B
A3 = B3	A2 = B2	A1 = B1	A0 = B0	L	H	H	L	H	H
A3 = B3	A2 = B2	A1 = B1	A0 = B0	H	L	H	H	L	H
A3 = B3	A2 = B2	A1 = B1	A0 = B0	H	H	H	H	H	H
A3 = B3	A2 = B2	A1 = B1	A0 = B0	H	H	L	H	H	L
A3 = B3	A2 = B2	A1 = B1	A0 = B0	L	L	L	L	L	L

H = high level, L = low level, X = irrelevant

86 $Y = A \oplus B = \bar{A}B + A\bar{B}$

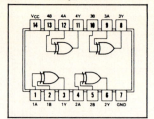

SN5486 (J, W)   SN486 (J, N)
SN54LS86 (J, W)   SN74LS86 (J, N)
SN54S86 (J, W)   SN74S86 (J, N)

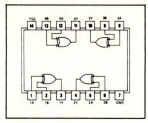

SN54L86 (J)   SN74L86 (J, N)

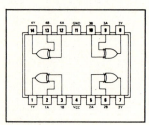

SN54L86 (T)

FUNCTION TABLE

INPUTS		OUTPUT
A	B	Y
L	L	L
L	H	H
H	L	H
H	H	L

H = high level, L = low level

64-BIT READ/WRITE MEMORIES

89 16 4-BIT WORDS

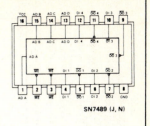

SN7489 (J, N)

DECADE COUNTERS

90 DIVIDE-BY-TWO AND DIVIDE-BY FIVE

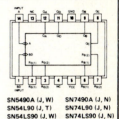

SN5490A (J, W)	SN7490A (J, N)
SN54L90 (J, T)	SN74L90 (J, N)
SN54LS90 (J, W)	SN74LS90 (J, N)

NC — No internal connection

BCD COUNT SEQUENCE
(See Note A)

COUNT	OUTPUT			
	Q_D	Q_C	Q_B	Q_A
0	L	L	L	L
1	L	L	L	H
2	L	L	H	L
3	L	L	H	H
4	L	H	L	L
5	L	H	L	H
6	L	H	H	L
7	L	H	H	H
8	H	L	L	L
9	H	L	L	H

BI-QUINARY (5-2)
(See Note B)

COUNT	OUTPUT			
	Q_A	Q_D	Q_C	Q_B
0	L	L	L	L
1	L	L	L	H
2	L	L	H	L
3	L	L	H	H
4	L	H	L	L
5	H	L	L	L
6	H	L	L	H
7	H	L	H	L
8	H	L	H	H
9	H	H	L	L

RESET/COUNT FUNCTION TABLE

RESET INPUTS				OUTPUT			
$R_{0(1)}$	$R_{0(2)}$	$R_{9(1)}$	$R_{9(2)}$	Q_D	Q_C	Q_B	Q_A
H	H	L	X	L	L	L	L
H	H	X	L	L	L	L	L
X	X	H	H	H	L	L	H
X	L	X	L	COUNT			
L	X	L	X	COUNT			
L	X	X	L	COUNT			
X	L	L	X	COUNT			

NOTES: A. Output Q_A is connected to input B for BCD count.
B. Output Q_D is connected to input A for bi-quinary count.
C. Output Q_A is connected to input B.
D. H = high level, L = low level, X = irrelevant

FUNCTION TABLE

INPUTS AT t_n		OUTPUTS AT t_{n+8}	
A	B	Q	$\overline{Q}_H$
H	H	H	L
L	X	L	H
X	L	L	H

H = high, L = low, X = irrelevant
t_n = Reference bit time, clock low
t_{n+8} = Bit time after 8 low-to-high clock transitions.

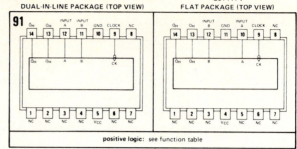

DUAL-IN-LINE PACKAGE (TOP VIEW)

FLAT PACKAGE (TOP VIEW)

positive logic: see function table

NC—No internal connection

functional block diagram

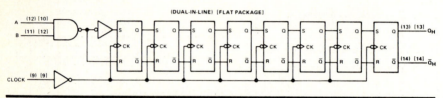

(DUAL-IN-LINE) [FLAT PACKAGE]

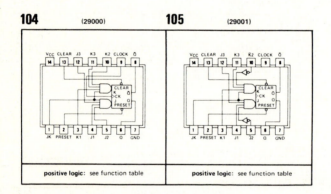

104 (29000)   **105** (29001)

positive logic: see function table   **positive logic:** see function table

FUNCTION TABLE

INPUTS						OUTPUTS	
PRESET	CLEAR	CLOCK	J	K	JK	Q	$\overline{Q}$
L	H	X	X	X	X	H	L
H	L	X	X	X	X	L	H
L	L	X	X	X	X	H*	H*
H	H	⊔	X	X	L	Q_0	$\overline{Q}_0$
H	H	⊔	L	L	L	Q_0	$\overline{Q}_0$
H	H	⊔	H	L	H	H	L
H	H	⊔	L	H	H	L	H
H	H	⊔	H	H	H	TOGGLE	

SN29000: J = J1·J2·J3; K = K1·K2·K3
SN29001: J = J1·$\overline{\text{J2}}$·J3; K = K1·$\overline{\text{K2}}$·K3
H = high level (steady state), L = low level (steady state), X = irrelevant
⊔ = low-level pulse; other inputs should be held constant while clock is low.
↑ = transition from low to high level
Q_0 = the level of Q before the indicated input conditions were established.
TOGGLE: Each output changes to the complement of its previous level on each active transition of the clock.
*This configuration is nonstable. That is, it will not persist when preset and clear inputs return to their inactive (high) level.

126
QUADRUPLE BUS BUFFER GATES WITH THREE-STATE OUTPUTS

positive logic:

Y = A

Output is off (disabled) when C is low.

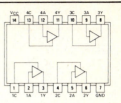

SN54126/SN74126(J, N, W)

147

(TOP VIEW)

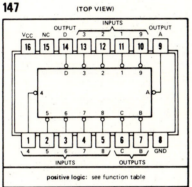

positive logic: see function table

NC—No internal connection

FUNCTION TABLE

INPUTS									OUTPUTS			
1	2	3	4	5	6	7	8	9	D	C	B	A
H	H	H	H	H	H	H	H	H	H	H	H	H
X	X	X	X	X	X	X	X	L	L	H	H	L
X	X	X	X	X	X	X	L	H	L	H	H	H
X	X	X	X	X	X	L	H	H	H	L	L	L
X	X	X	X	X	L	H	H	H	H	L	L	H
X	X	X	X	L	H	H	H	H	H	L	H	L
X	X	X	L	H	H	H	H	H	H	L	H	H
X	X	L	H	H	H	H	H	H	H	H	L	L
X	L	H	H	H	H	H	H	H	H	H	L	H
L	H	H	H	H	H	H	H	H	H	H	H	L

H = high logic level, L = low logic level, X = irrelevant

153

(TOP VIEW)

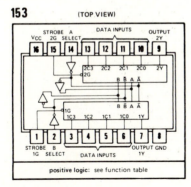

positive logic: see function table

FUNCTION TABLE

SELECT INPUTS		DATA INPUTS				STROBE	OUTPUT
B	A	C0	C1	C2	C3	G	Y
X	X	X	X	X	X	H	L
L	L	L	X	X	X	L	L
L	L	H	X	X	X	L	H
L	H	X	L	X	X	L	L
L	H	X	H	X	X	L	H
H	L	X	X	L	X	L	L
H	L	X	X	H	X	L	H
H	H	X	X	X	L	L	L
H	H	X	X	X	H	L	H

Select inputs A and B are common to both sections.
H = high level, L = low level, X = irrelevant

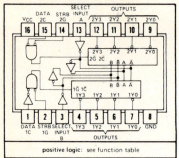

155

J OR N DUAL-IN-LINE OR
W FLAT PACKAGE (TOP VIEW)

positive logic: see function table

functional block diagram and logic

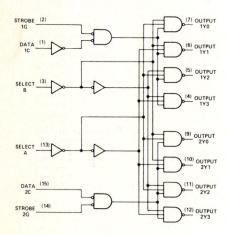

STROBE (2) 1G

DATA (1) 1C

SELECT (3) B

SELECT A (13)

DATA (15) 2C

STROBE (14) 2G

(7) OUTPUT 1Y0
(6) OUTPUT 1Y1
(5) OUTPUT 1Y2
(4) OUTPUT 1Y3
(9) OUTPUT 2Y0
(10) OUTPUT 2Y1
(11) OUTPUT 2Y2
(12) OUTPUT 2Y3

FUNCTION TABLES
2-LINE-TO-4-LINE DECODER
OR 1-LINE-TO-4-LINE DEMULTIPLEXER

INPUTS			OUTPUTS			
SELECT		STROBE	DATA			
B	A	1G	1C	1Y0	1Y1	1Y2	1Y3
X	X	H	X	H	H	H	H
L	L	L	H	L	H	H	H
L	H	L	H	H	L	H	H
H	L	L	H	H	H	L	H
H	H	L	H	H	H	H	L
X	X	X	L	H	H	H	H

INPUTS			OUTPUTS			
SELECT		STROBE	DATA			
B	A	2G	2C	2Y0	2Y1	2Y2	2Y3
X	X	H	X	H	H	H	H
L	L	L	L	L	H	H	H
L	H	L	L	H	L	H	H
H	L	L	L	H	H	L	H
H	H	L	L	H	H	H	L
X	X	X	H	H	H	H	H

FUNCTION TABLE
3-LINE-TO-8-LINE DECODER
OR 1-LINE-TO-8-LINE DEMULTIPLEXER

INPUTS			OUTPUTS							
SELECT			STROBE OR DATA	(0)	(1)	(2)	(3)	(4)	(5)	(6)	(7)
C†	B	A	G‡	2Y0	2Y1	2Y2	2Y3	1Y0	1Y1	1Y2	1Y3
X	X	X	H	H	H	H	H	H	H	H	H
L	L	L	L	L	H	H	H	H	H	H	H
L	L	H	L	H	L	H	H	H	H	H	H
L	H	L	L	H	H	L	H	H	H	H	H
L	H	H	L	H	H	H	L	H	H	H	H
H	L	L	L	H	H	H	H	L	H	H	H
H	L	H	L	H	H	H	H	H	L	H	H
H	H	L	L	H	H	H	H	H	H	L	H
H	H	H	L	H	H	H	H	H	H	H	L

†C = inputs 1C and 2C connected together
‡G = inputs 1G and 2G connected together
H = high level, L = low level, X = irrelevant

200

163

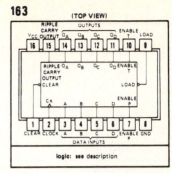

(TOP VIEW)

logic: see description

description

These synchronous, presettable counters feature an internal carry look-ahead for application in high-speed counting designs. The '160, '162, 'LS160A, 'LS162A, and 'S162 are decade counters and the '161, '163, 'LS161A, 'LS163A, and 'S163 are 4-bit binary counters. Synchronous operation is provided by having all flip-flops clocked simultaneously so that the outputs change coincident with each other when so instructed by the count-enable inputs and internal gating. This mode of operation eliminates the output counting spikes that are normally associated with asynchronous (ripple clock) counters. A buffered clock input triggers the four flip-flops on the rising (positive-going) edge of the clock input waveform.

These counters are fully programmable; that is, the outputs may be preset to either level. As presetting is synchronous, setting up a low level at the load input disables the counter and causes the outputs to agree with the setup data after the next clock pulse regardless of the levels of the enable inputs. Low-to-high transitions at the load input of the '160 thru '163 should be avoided when the clock is low if the enable inputs are high at or before the transition. This restriction is not applicable to the 'LS160A thru 'LS163A or 'S162 or 'S163 . The clear function for the '160, '161, 'LS160A, and 'LS161A is asynchronous and a low level at the clear input sets all four of the flip-flop outputs low regardless of the levels of clock, load, or enable inputs. The clear function for the '162, '163, 'LS162A, 'LS163A, 'S162, and 'S163 is synchronous and a low level at the clear input sets all four of the flip-flop outputs low after the next clock pulse, regardless of the levels of the enable inputs. This synchronous clear allows the count length to be modified easily as decoding the maximum count desired can be accomplished with one external NAND gate. The gate output is connected to the clear input to synchronously clear the counter to 0000 (LLLL). Low-to-high transitions at the clear input of the '162 and '163 should be avoided when the clock is low if the enable and load inputs are high at or before the transition.

The carry look-ahead circuitry provides for cascading counters for n-bit synchronous applications without additional gating. Instrumental in accomplishing this function are two count-enable inputs and a ripple carry output. Both count-enable inputs (P and T) must be high to count, and input T is fed forward to enable the ripple carry output. The ripple carry output thus enabled will produce a high-level output pulse with a duration approximately equal to the high-level portion of the Q_A output. This high-level overflow ripple carry pulse can be used to enable successive cascaded stages. High-to-low-level transitions at the enable P or T inputs of the '160 thru '163 should occur only when the clock input is high. Transitions at the enable P or T inputs of the 'LS160A thru 'LS163A or 'S162 and 'S163 are allowed regardless of the level of the clock input.

'LS160A thru 'LS163A, 'S162 and 'S163 feature a fully independent clock circuit. Changes at control inputs (enable P or T, or clear) that will modify the operating mode have no effect until clocking occurs. The function of the counter (whether enabled, disabled, loading, or counting) will be dictated solely by the conditions meeting the stable setup and hold times.

The 'LS160A thru 'LS163A are completely new designs. Compared to the original 'LS160 thru 'LS163, they feature 0-nanosecond minimum hold time and reduced input currents I_{IH} and I_{IL}.

201

170

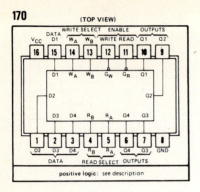

(TOP VIEW)

positive logic: see description

WRITE FUNCTION TABLE (SEE NOTES A, B, AND C)

WRITE INPUTS			WORD			
W_B	W_A	G_W	0	1	2	3
L	L	L	Q = D	Q_0	Q_0	Q_0
L	H	L	Q_0	Q = D	Q_0	Q_0
H	L	L	Q_0	Q_0	Q = D	Q_0
H	H	L	Q_0	Q_0	Q_0	Q = D
X	X	H	Q_0	Q_0	Q_0	Q_0

READ FUNCTION TABLE (SEE NOTES A AND D)

READ INPUTS			OUTPUTS			
R_B	R_A	G_R	Q1	Q2	Q3	Q4
L	L	L	W0B1	W0B2	W0B3	W0B4
L	H	L	W1B1	W1B2	W1B3	W1B4
H	L	L	W2B1	W2B2	W2B3	W2B4
H	H	L	W3B1	W3B2	W3B3	W3B4
X	X	H	H	H	H	H

NOTES:
A. H = high level, L = low level, X = irrelevant.
B. (Q = D) = The four selected internal flip-flop outputs will assume the states applied to the four external data inputs.
C. Q_0 = the level of Q before the indicated input conditions were established.
D. W0B1 = The first bit of word 0, etc.

FUNCTION TABLE

INPUTS			OUTPUTS	
Σ OF H's AT A THRU H	EVEN	ODD	Σ EVEN	Σ ODD
EVEN	H	L	H	L
ODD	H	L	L	H
EVEN	L	H	L	H
ODD	L	H	H	L
X	H	H	L	L
X	L	L	H	H

H = high level, L = low level, X = irrelevant

180

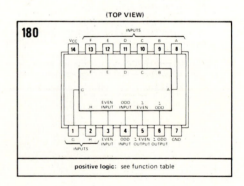

(TOP VIEW)

positive logic: see function table

(TOP VIEW)

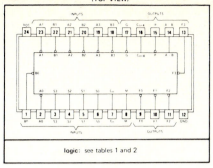

logic: see tables 1 and 2

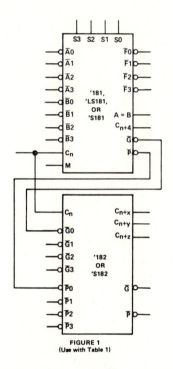

FIGURE 1
(Use with Table 1)

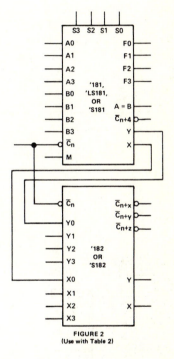

FIGURE 2
(Use with Table 2)

TABLE 1

SELECTION				M = H	M = L; ARITHMETIC OPERATIONS	
				ACTIVE-LOW DATA		
S3	S2	S1	S0	LOGIC FUNCTIONS	C_n = L (no carry)	C_n = H (with carry)
L	L	L	L	F = $\overline{A}$	F = A MINUS 1	F = A
L	L	L	H	F = $\overline{AB}$	F = AB MINUS 1	F = AB
L	L	H	L	F = $\overline{A}$ + B	F = $A\overline{B}$ MINUS 1	F = $A\overline{B}$
L	L	H	H	F = 1	F = MINUS 1 (2's COMP)	F = ZERO
L	H	L	L	F = $\overline{A + B}$	F = A PLUS (A + $\overline{B}$)	F = A PLUS (A + $\overline{B}$) PLUS 1
L	H	L	H	F = $\overline{B}$	F = AB PLUS (A + $\overline{B}$)	F = AB PLUS (A + $\overline{B}$) PLUS 1
L	H	H	L	F = A $\oplus$ B	F = A MINUS B MINUS 1	F = A MINUS B
L	H	H	H	F = A + $\overline{B}$	F = A + $\overline{B}$	F = (A + $\overline{B}$) PLUS 1
H	L	L	L	F = $\overline{A}B$	F = A PLUS (A + B)	F = A PLUS (A + B) PLUS 1
H	L	L	H	F = A $\oplus$ B	F = A PLUS B	F = A PLUS B PLUS 1
H	L	H	L	F = B	F = $A\overline{B}$ PLUS (A + B)	F = $A\overline{B}$ PLUS (A + B) PLUS 1
H	L	H	H	F = A + B	F = (A + B)	F = (A + B) PLUS 1
H	H	L	L	F = 0	F = A PLUS A*	F = A PLUS A PLUS 1
H	H	L	H	F = $A\overline{B}$	F = AB PLUS A	F = AB PLUS A PLUS 1
H	H	H	L	F = AB	F = $A\overline{B}$ PLUS A	F = $A\overline{B}$ PLUS A PLUS 1
H	H	H	H	F = A	F = A	F = A PLUS 1

TABLE 2

SELECTION				M = H	M = L; ARITHMETIC OPERATIONS	
				ACTIVE-HIGH DATA		
S3	S2	S1	S0	LOGIC FUNCTIONS	$\overline{C}_n$ = H (no carry)	$\overline{C}_n$ = L (with carry)
L	L	L	L	F = $\overline{A}$	F = A	F = A PLUS 1
L	L	L	H	F = $\overline{A + B}$	F = A + B	F = (A + B) PLUS 1
L	L	H	L	F = $\overline{A}B$	F = A + $\overline{B}$	F = (A + $\overline{B}$) PLUS 1
L	L	H	H	F = 0	F = MINUS 1 (2's COMPL)	F = ZERO
L	H	L	L	F = $\overline{AB}$	F = A PLUS $A\overline{B}$	F = A PLUS $A\overline{B}$ PLUS 1
L	H	L	H	F = $\overline{B}$	F = (A + B) PLUS $A\overline{B}$	F = (A + B) PLUS $A\overline{B}$ PLUS 1
L	H	H	L	F = A $\oplus$ B	F = A MINUS B MINUS 1	F = A MINUS B
L	H	H	H	F = $A\overline{B}$	F = $A\overline{B}$ MINUS 1	F = $A\overline{B}$
H	L	L	L	F = $\overline{A}$ + B	F = A PLUS AB	F = A PLUS AB PLUS 1
H	L	L	H	F = A $\oplus$ B	F = A PLUS B	F = A PLUS B PLUS 1
H	L	H	L	F = B	F = (A + $\overline{B}$) PLUS AB	F = (A + $\overline{B}$) PLUS AB PLUS 1
H	L	H	H	F = AB	F = AB MINUS 1	F = AB
H	H	L	L	F = 1	F = A PLUS A*	F = A PLUS A PLUS 1
H	H	L	H	F = A + $\overline{B}$	F = (A + B) PLUS A	F = (A + B) PLUS A PLUS 1
H	H	H	L	F = A + B	F = (A + $\overline{B}$) PLUS A	F = (A + $\overline{B}$) PLUS A PLUS 1
H	H	H	H	F = A	F = A MINUS 1	F = A

*Each bit is shifted to the next more significant position.

203

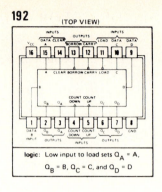

192

(TOP VIEW)

INPUTS
DATA CLEAR
V_{CC} A

OUTPUTS
BORROW CARRY

INPUTS
LOAD DATA DATA DATA
C   D

16 15 14 13 12 11 10 9

A   CLEAR BORROW CARRY LOAD   C

B   D

COUNT COUNT
DOWN UP

Q_B Q_A Q_C Q_D

1   2   3   4   5   6   7   8

DATA
B Q_B Q_A COUNT COUNT Q_C Q_D GND
INPUT   DOWN UP
     OUTPUTS   INPUTS   OUTPUTS

logic: Low input to load sets $Q_A = A$,
        $Q_B = B$, $Q_C = C$, and $Q_D = D$

194

(TOP VIEW)

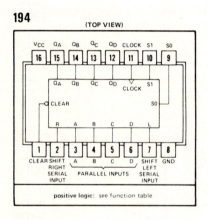

V_{CC} Q_A Q_B Q_C Q_D CLOCK S1   S0

16   15   14   13   12   11   10   9

Q_A Q_B Q_C Q_D

                        CLOCK   S1

CLEAR   S0

R   A   B   C   D   L

1   2   3   4   5   6   7   8

CLEAR SHIFT   A   B   C   D   SHIFT GND
      RIGHT   LEFT
      SERIAL   PARALLEL INPUTS SERIAL
      INPUT   INPUT

positive logic: see function table

FUNCTION TABLE

	INPUTS									OUTPUTS			
CLEAR	MODE		CLOCK	SERIAL		PARALLEL				Q_A	Q_B	Q_C	Q_D
	S_1	S_0		LEFT	RIGHT	A	B	C	D				
L	X	X	X	X	X	X	X	X	X	L	L	L	L
H	X	X	L	X	X	X	X	X	X	Q_{A0}	Q_{B0}	Q_{C0}	Q_{D0}
H	H	H	↑	X	X	a	b	c	d	a	b	c	d
H	L	H	↑	X	H	X	X	X	X	H	Q_{An}	Q_{Bn}	Q_{Cn}
H	L	H	↑	X	L	X	X	X	X	L	Q_{An}	Q_{Bn}	Q_{Cn}
H	H	L	↑	H	X	X	X	X	X	Q_{Bn}	Q_{Cn}	Q_{Dn}	H
H	H	L	↑	L	X	X	X	X	X	Q_{Bn}	Q_{Cn}	Q_{Dn}	L
H	L	L	X	X	X	X	X	X	X	Q_{A0}	Q_{B0}	Q_{C0}	Q_{D0}

H = high level (steady state)
L = low level (steady state)
X = irrelevant (any input, including transitions)
↑ = transition from low to high level
a, b, c, d = the level of steady-state input at inputs A, B, C, or D, respectively
Q_{A0}, Q_{B0}, Q_{C0}, Q_{D0} = the level of Q_A, Q_B, Q_C, or Q_D, respectively, before the indicated steady-state input conditions were established
Q_{An}, Q_{Bn}, Q_{Cn}, Q_{Dn} = the level of Q_A, Q_B, Q_C, Q_D, respectively, before the most-recent ↑ transition of the clock.

RESISTOR COLOR CODE

Color	Significant Figure	Decimal Multiplier	Tolerance (%)
black	0	1	
brown	1	10	
red	2	10^2	
orange	3	10^3	
yellow	4	10^4	
green	5	10^5	
blue	6	10^6	
violet	7	10^7	
gray	8	10^8	
white	9	10^9	
gold		10^{-1}	5
silver		10^{-2}	10
none			20

RECOMMENDED LED LAMPS

The cathode of discrete LED lamps is usually marked by a flat or notch on the plastic case.

Diameter (in)	Red	Yellow	Green
0.125	XC209R	XC209Y	XC209G
0.185	XC526R	XC526Y	XC526G
0.20	XC556R	XC556Y	XC556G

RECOMMENDED SWITCHES

Type	Style	Circuit
Dipswitch		
206-4	SPST	4 switch
206-7	SPST	7 switch
206-8	SPST	8 switch
Push button		
MS 102	momentary SPST	normally open
MS103	momentary SPST	normally closed

APPENDIX B
Parts List

PARTS LIST

Parts List in Numerical Order

TTL Integrated Circuits
(1 each except as noted)

7400 quad NAND

Experiments
1-4	7-1
2-9	7-2
3-4	7-7
3-5	8-3

7402 quad NOR

Experiment 2-10

7404 hex inverter

Experiments
4-4
10-7

7405 hex inverter

Experiments
2-7	6-1
2-8	6-5
3-6	7-1
3-8	9-3
3-9	

7408 quad AND

Experiments
2-4	3-8
2-5	3-9
2-6	6-1
2-8	6-2
3-2	6-3

7409 quad AND

Experiments
3-2
3-3
4-4

7410 triple NAND

Experiments
3-1
3-6
3-7

7411 triple AND

Experiments
2-5
5-1 (2)
5-2 (2)
8-4

7414 Schmitt trigger

Experiment 5-7

7423 quad NOR

Experiment 3-10

7430 NAND

Experiment 9-3

7432 quad OR

Experiments
2-4	4-3
2-6	5-1
2-8	6-2
3-8	6-3

7442 decoder

Experiment 5-5

7447 common-anode decoder/driver

Experiments
1-5
10-4
10-6

7448 common-cathode decoder/driver

Experiment 10-5

7474 dual *D* flip-flop

Experiment 7-5

7476 dual *J-K* flip-flop

Experiments
7-4
7-6
8-4 (2)

7482 adder

Experiments
6-4
6-5

7485 magnitude comparator

Experiment 5-6

7486 quad exclusive OR

Experiments
6-1
6-2
6-3

7489 RAM

Experiment 9-1

7490 decade counter

Experiments
1-5
8-3
10-6

7491 shift register

Experiment 8-2

74L71 flip-flop

Experiment 7-3

74126 quad buffer

Experiment 3-7

74147 priority encoder

Experiment 5-8

74153 dual multiplexer

Experiment 5-3

74155 demultiplexer

Experiment 5-4

74163 counter

Experiment 8-5

74170 register file

Experiment 9-2

74180 parity generator

Experiment 6-8

74181 ALU

Experiments
6-6
6-7

74192 counter

Experiment 8-6

74194 shift register

Experiment 8-1

CMOS Integrated Circuits
(1 each except as noted)

4049 hex inverter

Experiments
4-1
4-2
4-3

4071 quad OR

Experiments
4-1
4-4

4081 quad AND

Experiments
4-1
4-3

Other Integrated Circuits and Semiconductors
(1 each except as noted)

555 timer

Experiments
1-3
10-6

IN4001 diode

Experiment 1-1 (4)

IN456 diode

Experiment 5-7

IEE 1712 display

Experiments
1-5   *10-4*
10-2   *10-6*

MAN 74 display

Experiments
10-3
10-5

LM309K regulator

Experiment 1-1

LED

Experiments

1-2	7-4 (2)
2-1 (3)	7-5 (2)
2-2 (4)	7-6 (2)
2-3 (4)	8-1 (4)
5-8 (4)	8-2 (2)
6-1 (4)	8-3 (4)
6-4 (3)	8-4
6-5 (3)	8-5 (4)
6-6 (4)	8-6 (4)
6-7 (4)	9-1 (4)
6-8 (4)	9-2 (4)
7-1 (2)	9-3
7-2 (2)	10-1
7-3 (2)	10-7 (2)

Capacitors
(1 each except as noted)

62 pF mica

Experiment 1-3

0.001 μF disk

Experiment 1-3

10 μF

Experiment 10-6

10 μF tantalum

Experiment 1-3

2200 μF, 16-V, electrolytic

Experiment 1-1 (2)

Resistors
(1 each except as noted)

100-Ω potentiometer

Experiments
2-1
2-2 (2)
2-3 (2)

130 Ω

Experiments
2-1
2-2
2-3

15 kΩ

Experiments
1-3
10-6

1 kΩ

Experiments
4-4   10-1

2 kΩ

Experiments
9-1 (4)   9-2 (4)

330 Ω

Experiments

1-2	9-3
1-5 (7)	10-1
5-8 (4)	10-2 (6)
6-1 (2)	10-3 (6)
7-7 (2)	10-4 (7)
8-3 (6)	10-5 (7)
9-1 (4)	10-6 (7)
9-2 (4)	

4 kΩ

Experiments
2-7
2-8 (2)
3-2
3-3

68 kΩ

Experiments
1-3   10-6

900 Ω

Experiment 1-4 (2)

180 Ω

Experiment 10-7 (2)

Index